D0341346

SEASON
ON THE PLAIN

SEASON
ON THE PLAIN

FRANKLIN RUSSELL

READER'S DIGEST PRESS
Distributed by E. P. Dutton & Co. New York 1974

To Jacqueline

Library of Congress Cataloging in Publication Data

Russell, Franklin, 1926–
Season on the plain.

1. Animals, Legends and stories of. 2. Zoology—
Africa. I. Title.
QL791.R793   591.5'0967   73-20465

ISBN: 0-88349-024-2

SEASON
ON THE PLAIN

I

In those uncertain moments before dawn the island lay bulking and gray in the middle of an almost lunar landscape. Its rounded granite sides disappeared into dusky, enveloping shrubs; its crown sprouted triangular trees clipped by time into perfect geometric shapes. It appeared to move in the gloomy light, a solid mass of controlled energy awaiting escape from its union with the earth. But it would never move; its roots were so deeply buried it was a prisoner forever. In the long certitude of time, its fate was to drown in a sea of dust.

The night wind still blew, a steady westerly that had lost most of its moisture in its long journey across the high central African plains. Now, it carried dust whipped into cascades. The wind had dropped down mountain slopes, hurried through thorn- and scrubland, across lake and river and lightly treed savanna until it came to the plains on which the island was emplaced in its solitary grandeur. The dust that fell over the dried grasses was a part of this expansive country situated near the equatorial center of the continent, a place made benign despite the extremes of its moods by its height, nearly six thousand feet above the sea, and by its freedom from the humid oppressiveness of the distant coasts that flanked it to the east and west. The dust, breaking in granular waves against the island and rebounding toward where the sun would rise, was of a special kind, spawned by dead volcanoes.

The essence of the island was its aching aloneness, dumped in this infinity of flat earth encircling it, abandoned, an afterthought in some geological cataclysm that had meant to cover it completely. But the island had not been buried; it remained a fortress, an imperfection that reproached the symmetry of the plains.

Before the sun rose, and while the eastern sky and plains became suffused with gray light, the island did seem to be lost in lunar desolation. No animal could be seen moving between the island and the lightening eastern skies; no bird flew through a delicate band of pink growing above the sun's horizon. The island lay inert until the tip of the sun shone like an exploding star. Then, at this best moment of the African day, the hyraxes appeared. They placed themselves in precise positions, rodentlike in gray or brown fur and movements, but almost rabbitlike in appearance, except for their short ears, although by some cosmic joke they were related to elephants. One stood on a rock, another crouched against a boulder, a third climbed into a bush, a fourth stood on the branch of a tree. Once placed, they stood against the glare of the rising sun, watching. Nothing moved without being seen by the sentinel hyraxes. They were the island's eyes.

The appearance of the sun above the sharp line of the horizon washed the island in blood. It caught the silhouetted form of a gazelle grazing, a delicate figurine set in a frame of fire. The sun bolted upward and away and the blood-light washed down the island's rocks and was absorbed into the earth. The hyraxes watched, as they did every morning. Each new day was different from any other in its beginning. Sometimes they saw the horizon littered with antelopes resting, horses moving, giraffes grazing, lions copulating. Other mornings the plains were empty. But none of this meant anything to the hyraxes. Life eddied around their island, and most of it never involved them.

The sun climbed, and with it climbed the hyraxes, some to the tops of trees, others—the rock hyraxes—to the tips of rocks. They drove each other away from favored watching places, and the strong commandeered the safest positions. The tree hyraxes squabbled for the highest trees, although not with the venom displayed by the rock hyraxes in their desire for the best lower places.

In the rising sun the tree hyraxes went as high as possible to become observers of all traditional enemies: the arrival of a leopard, the occasional appearance of smaller cats—the servals, the grays, and the incredibly fast and cunning caracals. They watched for the falling attack of the great martial eagle, who came down with the sun behind him and might lunge at a hyrax's eye before being seen.

The hyraxes climbed with the sun, and the golden light

around them replaced the chill wind with a warming glow. Warmth was a reward. The hyraxes fought for the most sheltered crevices facing the sun. They squabbled for positions where they could enjoy the sun longest. Then, as the sun rose higher, they spread themselves out, limbs limp. In places they were packed so closely together they looked like a composite fur animal covering the rocks. Youngsters climbed from one adult back to another.

After the sun became white and hot, the island rocks absorbed its heat and held it, and the hyraxes moved, seeking shade now. Yawning, they started to feed, to groom, to ready themselves for work.

The island, long deprived of rain, had been so well eaten over that hardly any food remained at earth and rock level. To survive, the rock hyraxes had been forced to change their night-feeding habits and their close dependence on the shelter of the island, leaving its refuge and so becoming reliant on the sentinel tree hyraxes to warn them of danger. Although their low positions in a flat world made them sensitive to the minute sounds and smells of danger, they were not well matched against the dissimulative approach of the jackal or the stalking fast charge at dusk of the caracal or serval. When they went out to the plains, they did not know whether an enemy had ensconced himself on the island during the night; they had no way to determine at what point they would outdistance their capacity to return to the island if an enemy were present there.

And of course, as long as there were hyraxes there would be enemies. On one rounded rock lay the coiled figure of a sun-basking python, thirty feet of animus which contained the digesting body of an antelope. The python was neutralized for the moment by the huge meal. In the island's deepest cave a striped hyena dozed, herself hyrax-prone, but not dangerous until dusk, when she would come out cautiously to hunt. Neither python nor hyena, however, could influence the behavior of the hyraxes who were accustomed to such relatively mild dangers. But baffling this morning was that no hyrax eye had probed the branches of one stunted tree set among rocks on the northern shore of the island. The tree looked empty enough; yet it hid the long form of a leopard, his spotted hide breaking up the intelligence of vision and fooling it for a second into believing he was not there. He lay limp as cloth, both back legs dangling down in silent contradiction of the grace and power and fury quiet within him. He lay in what seemed to be such utter relaxation, such apparent contentment that no visible hint existed that he himself was a refugee from the plains, an alien on the island. The leopard was an animal of cover, continuous cover throughout his range of hunting. Without cover he could not use his techniques of the silent drop from the high branch, the ferociously powerful spring from out of the bushes, the blurringly fast sprint to claw down a running victim. He, the strongest cat on earth for his size, was so much developed to short explosive action that all open spaces were hostile to him. The plains demanded abilities he did not possess—endurance, gregariousness, the capacity to dig and to endure hours of direct sun, the stamina to move a hundred miles in a night to cope with some unexpected change in his territory.

His true country lay so far away that it was an unresolved dream, a desolation of distance separating him from it. A long line of trees flanked a watercourse there, thickening in swampy lands or clustering forests, skirting hills where scrub sprouted, surrounding ponds or small lakes or springs where animals drank. There, he could range for miles in two directions and make the territory entirely his own, driving other leopards from it. But there, one terrible day, he had collided with the inflexible system of selection in the form of a more cunning leopard who had surprised him, who had fought him uncharacteristically, who had crippled him so badly that his effort to resolve the fight had been focused on escape. Flight, blind flight, had brought him into immediate collision with the leopard in the next territory. Then, two young lions had chased him, driving him into the dreaded open. From then on, a blur of hysterical days was shrouded in pain and punctuated by angry lions, inquisitive hyenas, and the stench of hunting dogs.

He had fled, crippled still, into territory now permanently alien, his flight turned into sustained panic in which all his guile and cunning had been obliterated. The dawning of each day in country sometimes empty of even a solitary tree had nourished the horror of the night. Late-flying owls passed overhead, and ostriches watched him from nests placed absurdly in the middle of nowhere. He did not know where he was. The nightmare had been prolonged because the blood from his wounds was scented by dogs. He had reached a lone rock and defended himself there against a ring of enemies until they tired of him.

Finally, he had found refuge on this island. But it was refuge consisting only of the slender cover of the island's body, and it offered him no hope of surviving there long.

It was so absurdly limited compared with his needs that his power was diminished by comparison. Here, he was a victim of the implacable landscape surrounding him. The golden flatness was broken only by the hump of an occasional small island and the branches of stunted solitary

6

trees. He could not move anywhere without coming into open country where, he knew, luck could become disaster under the praetorian rule of packs of dogs by day, tribes of hyenas by night, prides of lions both day and night. He was trapped on an island where his favored victims did not seek cover. No small antelopes minced under his tree; no troops of monkeys arrived to cavort and eat seeds and fruit in the trees around him; no cranes nested in the flat land nearby. No island tree was high enough or strong enough for him to store his larger victims. Instead, he sprawled on the thin branch of his stunted tree in poor enough conceal-ment that would have to endure until his wounds had healed. His somnolence, his sprawl of unconcern said noth-ing about what had brought him to the island. He slept in recuperation, gathering strength and the determination to return to the place from whence he had come, where he truly belonged.

The hyraxes belonged in this place among the hoary rocks and dried sprouts of grass, the soaring sculptures of stone and squat thornbushes, and with the habit of millen-nia ingrained, they set about their work. The tree hyraxes reached their browsing positions long before the rock hy-raxes had ventured far from their refuges, the urgency of their demeanor suggesting the brevity of their time. Like large rats they slid away from the island into the empty, open spaces. All the available fodder, the dried grass and creeping, minuscule shoots of green had long since been eaten out around the island. They were forced to venture further and further each day. In an effort to minimize the danger of attack, they spread wide. But nervous, they turned frequently to face the island, making sudden rabbit-runs back toward it. The sun was not near its zenith before both rock and tree hyraxes began their retreats to the rocks once more.

The leopard appeared to be asleep, but like all cats he rarely slept deeply. When the hyraxes first began to feed, one of his eyes opened occasionally to track possible victims, while his brain recorded changing positions, collating the information even when both eyes were closed and he seemed soundly asleep. It was only when the retreat from treetop and plains had started that he moved. His place of concealment had yielded itself to no watcher. Despite the pain of his injuries, he slid down the shadowed side of the tree like dappled water, his body dissolving into the ground and disappearing among the rocks. His pain was subordinated to his intuitive capacity to move so quickly, so deftly that his victims might not suspect he had even stirred. Thus, he was in an attacking position on the plains before any of the hyraxes had a chance to whistle alarm.

Years of hunting, eons of race experience told him this was the best time to attack, while the tree hyraxes were preoccupied with their task of returning to earth. The leopard was actually a dozen steps beyond the shores of the island before the first whistle of danger sounded. By then it was too late.

The rock hyraxes did not, could not, hesitate. No choices were offered. Every one of them knew it was futile to run deep into the empty plains. They must return to the island, and they could do it only one way. They charged the leopard. Small, furry creatures, their tiny feet scampering, in extremis they revealed nothing of their elephant ancestry. The leopard slowed his forward run so that he could select his victims sensibly. With three deft movements of his right and left paws he broke the backs of three hyraxes. He could have killed more but did not. The whistles from the treetops pierced his ears, but he ignored these futile cries and gathered the dead in his mouth and returned to the island. There, all was confusion. In panic the hyraxes collided as they sought desperately to reach any refuge. Several leaped into alien burrows, there to meet cobras or puff adders, and one died in the jaws of the aroused striped hyena.

Slowly, the confusion subsided, the leopard became invisible again, camouflaged along the line of his branch, and no sign of life showed anywhere. The two uneaten hyraxes were jammed in a fork of the tree, mouths agape and dead eyes gleaming, as a solitary martial eagle, looking down for tree hyraxes he might panic into runs for safety, saw only a stillness of stone.

The island again appeared to be lost in a landscape of the dead. The wind had changed and was now hot and dry from the north. The sun caught no shadows on the plains. Now overhead, it penetrated the foliage of every plant on the island. It punched heat into every crevice, felt its way into rifts and caves and narrow holes. Its oppressive insistence caused the plains to expand their aspect slightly. The early light had been deceptive. The once-empty landscape revealed itself as not empty at all. Faint blue hills crouched against bunched mountains of clouds in the south. A series of islands protected a well-treed, permanent water hole in the north. The rounded tip of another island rose from the western horizon. East of the island the inevitability of flatness broke itself briefly in scattered trees and a lengthening of grasses around a declivity where moisture lingered and where, when rains came, a water hole would be reborn and survive long into the next drought.

The dust, eddying, rushing, disappearing, was the manifestation of the long drought—two hundred days—but the plants of the island, bunched together in a kind of comradely union against the pressure of the empty plains, told of abundant rain. Yet rain falling from such a sky, impassively blue and white, the sun staring, was contradiction. The notion of rain, difficult enough to conceive, was no more incongruous than the mystery of the serene landscape itself, as parklike and orderly as an architect's mind. It was so pure, so stark in its simplicity and grandeur that it suggested innocence; yet violence lurked in it like the absent rain. Anticipation hung in the dry hot air, was present in the shape of the bulking clouds, hid among the shrubs and trees of the island, lay uneasily in the shadows of rocks. It lurked in the quality of the light itself, which was unique, a special product of this high country set so near the equator. Perhaps the rain did not, in fact, fall; perhaps the island and the plains were a chimera, the plants fossils, products of a fossil rain.

The sun turned down into the west, threw diagonal light, and the face of the plains and the lonely island changed again. A small group of gazelles, husky yet graceful dun-colored antelopes who made these African grasslands their home in all seasons, moved impatiently forward for mouthsful of invisible food hidden in the dust. Their heads came up suddenly and sharp, slender horns turned interrogatively toward possible danger. Previously, they had been camouflaged against the golden background of dried grass and bare brown earth. A lone ostrich stood as motionless as a tree. A falcon, poised in midair, was held in the eternal wind, his blunt head turned downward. On the plains, dust whirled and a black beetle bumped past the gaping hole of a spider. A black road of ants cut through the debris of dead grasses, twin lanes packed with hurrying creatures ducking into black holes to bypass footprints, then branching into a maze of other roads and tunnels.

Deception once more. From whence had come the suggestion of a lifeless landscape? The sun moved more quickly downward, and hyraxes appeared again, memory of the leopard attack faded. Life on the island would have been impossible if it were based on the total recall of all the disasters that had occurred. Hyrax eyes resumed their work, observing the fall of a feather, the stagger of a beetle, the float of a wasp, the abrupt flight of a single locust one mile distant. Hyrax eyes saw tiny flies and midges whisking to and from the island, their wings making notes in musical scales carried on the wind. Larger insects whined and zipped by, only heard by most creatures but seen as well by the watchful hyraxes.

They recognized a blurring black-and-white movement to the south; it was the quick flight of a lapwing. They saw a silent clot of lovebirds packed close and moving fast toward the east. There was information in the high soar of an eagle, near invisible to most eyes. The hyraxes could distinguish him from a vulture, even though distance diminished the eagle to the significance of a fly. Information was everywhere, every second, but was available only to those, like the hyraxes, who could read it correctly.

They watched as rounded ears and dark heads emerged from termite mounds. The bat-eared foxes moved quickly to a second mound. From another mound, worn almost flat by time, came a clump of mongooses tightened into a fighting ball which flipped quickly through the dust and disappeared into the earth again. Hyrax eyes examined every detail of their island world. This time they correctly located the sprawled body of the leopard, his camouflage momentarily made absurd by the inquiring chatter of a sunbird, iridescent green plumage glittering in the brilliant light as she hovered at the sleeping cat's muzzle. The leopard had eaten the second hyrax and now lay with both back legs hanging down on either side of the branch, his fore-limbs stretched ahead of his muzzle, eyes closed. But with the caution of his kind he listened as closely as did the hyraxes themselves. Every sound passed into his brain, giving him a report on the condition of the day world beyond his shadowed retreat.

The leopard's eyes remained shut to the cry of the eagle who, observing the hyraxes, had come closer to the island. High, remote conversation from a dense-packed mass of streaming blackbirds did not stir him. Standing in a nearby tree, the late sun glinting on his purplish-blue feathers, a starling intoned one of his many extraordinary statements. His beautiful voice warbled pure music, then interrupted the song for a moment to inject a metallic sound, and concluded with an inquiring whistle. Two barbets, stocky yellowish-gray birds, stood on a branch and sang a duet, bobbing and bowing, a real clashing of voices until one of them, the disharmony unbelievable, raced away.

Although his eyes were closed, the leopard could place his vision anywhere on the island. He knew every part of it with an intimacy that only a great hunter would need. It was his fortress in a country so poorly covered that every rock or tree or shrub or patch of long grass was precious to his safety. He had sheltered in many such islands, though this one seemed more solitary than most because it was so far out on the plains. The shapes of the islands were similar, yet never quite the same. They had survived burial under rains of volcanic ash which had created the flatness of the plains originally. The leopard had learned to recognize many of them. He could associate the shapes of some islands with the kinds of inhabitants they sheltered. He had known one near the woodlands country, rounded and small, which harbored a serval cat, a civet, and a hyena. On the horizon, a larger island with towering pinnacles of tree-clad rocks had held a caracal, a score of hyraxes, a vulture roosting place, an eagle's nest, and countless smaller animals. He had visited another island where every two years a leopard bore her cubs. He knew islands where cheetahs took shelter to give birth and one island where hunting dogs made their headquarters every year or so.

On this island, he knew exactly the relative placement of every boulder, every tree and stone. In one place jumbled rocks formed a cave big enough to harbor lions or civets and genets, foxes and porcupines. Deep inside the cave bats hung in sleep. Wasp daubings marked the ceilings and walls. Paper nests tossed in random winds and skulls gleamed in late afternoon glows of sun. But he was no cave dweller.

From his tree at the end of the island he could, without opening his eyes, visualize the great slabs of granite flung down in disorderly memory of ancient ages, weathered and split and smashed by the expansions and contractions of cold nights and hot days. In the midmornings, in particular, he had learned to control the sharp thrill of fear that made his body shudder when one of the rocks, warmed by the sun, exploded away from the grip of the chill night.

None of this was dangerous, though it was useful in ways that hardly concerned the leopard. The almost constant breaking up of the rocks manufactured tiny caves and crevices where new lives could establish themselves. The rocks frequently were arranged so that they caught and channeled the rain, feeding it into the body of the island. This gave many plants the chance of dense growth and created a variety of humidities, the degrees of dampness graduated precisely through a whole range of life opportunities, from tiny bugs to giant snakes.

The leopard lay in one of these environments, albeit well separated from the body of the island, with shade above and cooling winds below. In different ways other animals found cool places in the island during the heat of the drought. Simultaneously, such places were warm during the sometimes cold nights. The brittle, dry air of the drought was tempered by water released grudgingly from the center of the island. But when the outside plains world hissed with torrential rains, some worlds inside the island were drier because the island threw off the water it did not need.

With the leopard's eyes shut, hyraxes watched the gentle fall of the eagle, the upright, motionless ostrich, and their eyes magnified the vision of the plains into a tumult of activity. They saw ants scrambling along every twig of bushes, flies darting about the remains of last year's fruit.

The sun accelerated its fall and so increased the anticipation, or apprehension, of coming events. Hyrax eyes had seen the shuttling images of countless thousands of other days, of other years, of dust puffs rising from each heavy-footed step of solitary lions, of drab pipits and chats whipping along in search of insects, of bright legions of finches dropping mosaics of color among seeding grasses, of motionless lizards and angular bustards which looked like storks but had vulturine appetites, of creeping quails and bustling termites, of hunting spiders and marching ants, of larks and storks and curlews and mice, of hares and grasshoppers and bugs and beetles, of butterflies and hiding moths and vultures that used stones to break ostrich eggs, of small hawks that watched larks catch insects and pirated them while larger hawks watched the smaller hawks, of eagles that watched hawks and eagles that watched eagles.

One day soon the hyrax fortress would be flooded in a sea of animals mooing and braying and hooting past the island in day-and-night parades. These animals would bring with them the cats and the dogs, the hyenas and the eagles, the vultures and the meat-eating storks. The hyraxes had seen the high, vertical, useless jump of gazelles, had heard the thud of a lion's paw striking, had listened in the night to the screams of stricken zebras, had felt the pounding of thousands of feet fleeing from night enemies. But now, despite the leopard in his tree and the single watching eagle, they were as safe as they would ever be.

The sun touched the horizon with a band of pink, and the island darkened in response. One hyrax darted from the rocks and ran a score of paces out onto the plains and stood up to watch. Nothing. A second hyrax joined him and stood up. Nothing. The leopard remained still; the eagle had swung away to the south. Another group of hyraxes appeared from the rocks and scaled trees around the leopard.

The python, feeling the warmth of the rock diminishing slightly, stirred herself and flowed out of sight among the boulders. The leopard opened his eyes and looked directly into the sun. In the distance the grumble of a lion sounded like small thunder. Now, the hyraxes played a deadly game, matching the distance they could prudently move from the nearest refuge against the leopard's capacity to cut them off. They could not know that he was not especially hungry, but perhaps they intuitively suspected it from his slow response to the dying day. Certainly, they were engrossed with the leopard. Their every action was satellite activity around the lethal heat of his presence. Thus, while they scampered and stood, grazed and ran, they did not see that another enemy was on the island. None of their eyes were sharp enough to catch the caracal; his tawny body and black tufted ears merged perfectly into the background of the island.

The sun dropped, and down with it came the vultures, a score of them from all parts of the sky. They circled widely around the island and then collapsed swiftly to it, the rushing of their wings sounding like falling rain. They settled, rustling silken wings, in the highest branches of the trees. None of the hyraxes moved. The leopard ignored them. The sun dipped into the horizon and was cut in half by the black line. The sky exploded scarlet, revealing a thin string of black clouds. An owl, unseen until this moment, left the island and flapped ponderously away across the darkening plains.

The rock hyraxes began bolting back to the island even as the caracal made her move. She intercepted one of them and whisked the body away so swiftly that few of the hyraxes knew one of their number had died.

Meantime, small birds came discreetly out of the plains and disappeared into foliage. Flies, butterflies, bees, and wasps drifted in and concealed themselves. Mounting anticipation defied the realities of the day. It was clear that the island was also a night place and that much of its mystery would be revealed under moonlight rather than in the inhibiting fire of the sun.

The leopard stood and yawned, the hyena came fully out of her cave, and the far distance revealed a solitary black running figure. It scampered, turning occasionally to stand up, manlike, and look behind. The leopard saw it immediately and tensed, golden eyes wide at the proximity of his favorite victim. Behind the black figure a trio of hyenas were strung out in loping, unhurried pursuit. But each time they closed the black figure turned and stood up, appearing to double his size. And each time, the hyenas, uncertain, veered away.

In this way the baboon and the hyenas came to the island. The leopard dropped to a place of ambush. The baboon approached the island on its northern side, where it was dominated by almost vertical rock cliffs. In one last desperate rush, he reached these cliffs, the hyenas only bounds behind him, and jumped, arms swinging. He secured one handhold as the other broke away rock. The hyenas leaped for his genitals. Another lucky handhold drew him up rapidly, feet scrabbling. Another handhold. He was a black ball of desperation fastened to the rock wall. The excited yips of the hyenas sounded beneath. But he got another handhold, the light fading fast now and the cries of hyenas diminishing as he disappeared over an overhanging ledge. He faced another almost vertical rock climb, but the rock was riven with cracks and there were handholds aplenty. He began climbing. Here, at last, was refuge; here, a place where he could pause and let the memory of his flight across the plains fade from his brain.

The journey to the island had started in such utter surprise that the baboon would never forget its beginning. As a senior member of a small troop of his fellows, he had been a totally gregarious creature. So, when a leopard had attacked the troop one night, he had remained at the top of the refuge tree, holding himself firmly against the swaying branches and listening to the scrabbling of feet and the screams of his terrified comrades. He had heard the thump of bodies hitting earth, then gradually the sounds of the attack had diminished until silence had smothered the night once more. The leopard had gone.

But so had the other baboons. When dawn came, he had been alone. Worse, when he had attempted to let himself down to the ground, a pride of lions had appeared among the trees and sprawled around his refuge, bellies bulging with meat. He had remained there for three days, not knowing that a second, and then a third leopard attack had so decimated his troop that the disorganized and disheartened troop elders had migrated in search of safer living grounds.

Alone, the baboon had not been certain of any kind of security at night. A young leopard had marked him for the kill and had attacked him every night until the cat had stuck a claw into one of the baboon's hands, ripping it open to the bone, and hurled him into darkness.

He had fled from the leopard, his escape horrendous in open country. Expecting to find trees, he had found none. He had run beyond the range of the leopard, but had entered the country of the hyena and dog, the lion and jackal. He had crouched in the open entrances of aardvark burrows and sheltered in the dens of bat-eared foxes.

Darkness rushed across the plains and enveloped the island in a tide of night. The baboon climbed confidently now. The sight of the tallest tree on the island was lodged in his brain. He climbed toward the curving crest of the rock cliff where, flat as a python's body and scarcely breathing, the leopard waited on rocks still warm from the day.

II

The roots of the island penetrated a billion years into rock, solid crystalline rock which underlay most of the continent. Its endurance was measured through countless years of erosion, of changing temperatures and climates, of the advances and retreats of mammals and vegetation, insects and reptiles and birds. It had hosted men and premen, and others who came after apes but before the premen. It had stood there while the plains around it rose in storms of smoke and blizzards of falling ash as nearby volcanoes erupted. Most of its geological history was buried beneath the level of the plains in caves filled in eons before, and its endurance indicated time too prolonged to be measured sensibly. For thousands of years no eruptions had struck, and the plains, no longer replenished by them, were being carried away, particle by particle of soil moving an inch or two each year toward meetings with rivers and lakes which took them into distant oceans and seas.

The island stood as a monument to the length of evolution and the quality of survival. All life, whether on the plains or in the woodlands, measured itself against the commanding island. It remained conservative, relatively changeless, stoic and immovable in a movable world. Yet its success was achieved through only tiny adjustments, minuscule variations from the norms of life. Its stable humidities and temperatures showed this. While the hyraxes dozed away the middays of drought in rock refuges where the air temperature was eighty degrees, the creatures of the plains moved in temperatures of one hundred and twenty degrees. When cold western winds cut across the midnight plains, the hyraxes slept in seventy-degree nights while the creatures of the plains hunted, or were hunted, in temperatures of forty or lower.

This equability of the island, its capacity to level the peaks and valleys of stress, sharply contrasted with life on the plains and beyond. The antelopes still roamed widely, many of them able to extract moisture from the grasses they ate, but the great meat eaters were becoming more dependent on the water in slowly drying water holes. The island lay serene and cool and waiting while chaos ruled these water holes. Irascible elephants gathered and drove off thirsty lions. Rhinoceroses were pushed aside by the elephants, tossed into shrubbery and thrown into the muddied water. The water receded; all feeling for sharing it disappeared. Elephants drank, then wallowed the water into undrinkable soup. Zebras gathered, poured into the water, stomped it into mud, then moved onward.

The island had stored water when the rains had fallen. This enabled some shrubs to reach down long tubers and thrive throughout the drought. Elsewhere, when thirsty elephants came on these shrubs, they dug them up and sucked all the moisture stored in the long white roots. The hyraxes were not able to reach water now, but they passed the moisture of leaves and tiny sprouts of green grasses so slowly through their bodies that their urine was as thick as sap, dark brown and viscous.

The baboon, his long arm stretched ahead in the gloom for another grip on the shelving rock, had not drunk for days either, but he too had stored moisture in his body. He had obtained the moisture from innumerable, almost invisible sprouts of green at the bases of some grasses. They seemed quite dead in the drought but they kept one tiny point of green alive, waiting for rain to fall again. Theirs was the ultimate distillation of the remnant moisture left in the universe of the plains, and the baboon survived on it.

No adjustment to the drought was haphazard or even recently evolved; it was an extension of very old arrangements. The smallest details of survival had been refined to subtleties too small to measure. Each tree and shrub and grass and living thing had made such individual adaptations so that each displayed a separate personality.

Some members of the acacia family—the tortilis—had broken away from family tradition to flower and grow new leaves during the drought. This was possible because they shared at least one of the island's secrets. Their long taproots reached down to drink water stored there. A relative tree had developed a slight variation by losing its leaves during the previous rains, only to burst into leaves and flowers when the drought began. The acacias, so long refined into this world of tricky variations of climate and conditions, equaled the island in their capacity to outlast the drought, even profit from it. Some were capable of anticipating the rains so precisely that they would break into green leaves only days before the rains began.

The island hosted a thousand plants, but there was no agreement among them about how best to handle the drought. Many had developed their resistance to it before they came to the island. The commiphora trees were deciduous and would grow quickly in the rains, although now they had dropped their leaves and were resting. The combretums, a clustering family of shrubs, had compromised to both the season and the heavy browsing. Some had shed their leaves so they could rest during the drought time. Others had remained evergreen and grown multiple slender stems to combat the restless, searching mouths of browsing animals.

In his flight to reach the island, the baboon had passed over a mosaic of grasses fighting the drought with individual devices. Their positions on the plains had been determined long ago after the showers of volcanic ash. The ash had not fallen evenly or homogeneously. It was concentrated in different types, depending on the kind of eruption, the direction of the wind, and the composition of matter expelled. Each grass fitted into the area most finely suited for it or had developed qualities to exploit the area.

In his haste the baboon had passed food in places where light touches of rain had stimulated some grasses to premature growth. The island's great bulk was its insurance against the vagaries of such stimulation. Nowhere on the plains did the rain fall evenly or predictably. None of the plains dwellers could ever be sure what profile the season would present. While the island stood in drought, thirty square miles of distant grassland were getting an inch of rain, a comparative torrent when matched against two hundred days of drought. But for a thousand square miles around the rain the drought continued. Animals who had turned to this ephemeral area of plenty quickly found themselves isolated there, the grass dried out, the temporary water pools disappearing overnight.

Then another heavy shower twenty miles away brought up a burst of grass, and the antelopes turned toward it. Advances and retreats, hungers and surfeits governed the movements of the animals on the plains, while the island sat in lonely magnificence, unaffected by such ephemera. The island's capacity to outwait disaster, to spread abundance, to remain equable while the world beyond its boundaries fluctuated, was imitated by the lowliest forms of life. Many of the grasses around the island had made adjustments to the violent extremes of weather. Now, they had dried completely, disappearing in some places so that only dusty bare earth remained where they had grown. Their shriveled bodies had been cropped by eager antelopes, picked over by the prowling hyraxes, carried away by industrious termites.

These grasses waited, quiescent, their ability to remain intact as living entities preserved because they grew from the bases of their leaves, not from the tips as did most other plants. They were thus ready to spring to life again the moment the rains came. If they were stimulated into growth and grazed down by passing animals, they were able to resume growth moments after their leaves had been eaten. The baboon had scurried across such grasses in his long and frenzied journey over the open plains.

During the early part of his flight he had traveled in fire country, where longer grasses, fringing the true open plains, had burned recently. He had not delayed there because there was nothing to eat except the odd charred crisps of insects overtaken in the flames. But in this country another kind of grass, the red oat, had devised a way to survive both drought and fire. This grass took an ingenious shortcut to get its kind perpetuated as rapidly as possible. It produced seeds with backward-facing tails, or bristles. Thrust into the ground, these seeds expanded and contracted with the heat of the sun and the moisture of night dews and light rains. This action twisted the seeds deeply enough into the soil to protect them from fire and ensure germination.

The baboon, a relatively primitive primate, was a vegetarian with an appetite for meat, and so seemed to be an evolving animal ascending to higher levels of opportunity. He ate tree seeds, buds, grass shoots, fruit, but he also ate young birds, eggs, hares, and newborn antelope. In his journey to the island he had been driven on by a hungry and hostile landscape that bristled with alarms. Everything was alien. His sense of the foreign, the unexpected, and the terrifying pushed him along in sustaining ignorance of the world around him. Thus, he passed opportunities to pause and refresh himself. He rushed, terrified, along the fringe of a dried-out watercourse, the smell of cat strong everywhere, and knew nothing of the plants that grew underfoot and were so well habituated to drought that their survival was assured by long tubers which reached down to residual waters in the earth. When these waters ran dry, the plants stored what they had gathered in corms, in bulbs, in long, widely spread roots. He had run across short saw-toothed grasses which thrust up seeding stalks as high as possible while keeping their leaves flat on the ground so they could resist the ceaseless grazing. When their seeding stalks were eaten off, the flat-lying leaves remained untouched and the plants recovered quickly, thrusting up other seed heads within days. At the same time, they produced runners under the surface which would, in time, make new, quick-growing stems that would shoot upward the moment the grazing animals had moved on. New seeds fell to the ground only days after the grazers had gone.

The baboon had run unheeding among lilies which looked dead, their dried foliage flattened on the dusty earth. But their conservation of energy in deeply buried roots was so complete they would be ready to begin growing within hours of a rainfall. Unlike many of the other plants which would store water as soon as the rains began as insurance against drought that sometimes followed the early rains, the lilies would stake all in a rush of growth. They could put up stems, flower, and seed within a few days under the sponsorship of short, light rains. If the rains continued, they then stored water in their roots, but by that time their seeds were already in the ground, perhaps even sprouted, as they made ready to face drought.

The baboon was now on top of the rock. He halted, his eyes penetrating the fading light as he hunted for the high tree he would scale and examined the terrain for sign of an enemy. The flat leopard, the barely breathing leopard, the patient leopard—his moment had come at last after a trembling wait of soft scratches and strong baboon smells. He hunched himself slightly, the forepart of his body raised on muscles that ran the length of his body. The right paw poised for the scimitar slash that would pin or disembowel the baboon, the left paw tensed, the reserve weapon to hold the stricken animal if the first blow failed to kill.

Because of the extremes of drought and wetness opportunism ruled the new world of the refugee baboon. Life had to be as nimble in its manifold ways as the baboon's startling climb up vertical rock, as quick to exploit as the leopard's rush to his place of ambush. The nimble and the opportunistic united in a sum of survival. And opportunism was everywhere, not just among the perennial plants or in the leopard's patience. The annual plants grew afresh each year from seeds, their parents always dying within the year. The seeds could lie in the dusty ground for months, even years, and remain viable. With rain, they rushed to produce flowers. Some were placed so low they were beneath the reach of a gazelle's teeth. The plants timed their rush to seed in order to beat the grazing animals. Grazing stopped them from seeding and trampling might wipe out all the annuals over large areas of the plains in bad years.

Some of the annuals were so finely adapted that after the initial flush of rain they were nothing more than a single stem thrusting up a single flower. But if the rains continued some of them made contributions to the wealth of the grasslands by growing masses of leaves that were good for grazing. In truly heavy rains which flooded wide areas of the plains in some years, these tiny, obscure annuals with their diffident flowers could become plants more than a foot high, long enough and lush enough to fatten a hundred thousand zebras.

The annuals had to be opportunistic. Most of the territory of the plains consisted of a mosaic pattern of perennials. The annuals were thus limited to the temporarily

27

unexploited areas that ran between the communities of perennials or to the bare patches created by overgrazing, trampling, or flooding.

In many direct and indirect ways the welfare of the island, of leopard and baboon, of all the grazing animals and migrants hinged on the obscure permutations of the humble grasses. The complexity of their world was beyond any simple statement. Deepening the mystery, they were not consistent in their behavior and so put higher forms of life at their mercy. A second generation of annuals might flourish during the long rains but, conversely, it might not appear at all. The plants seemed possessed of an awareness of events to come and knew, as it were, when it would be useless to spawn a generation with no chance of survival. They might be suppressed by more successful competitors and so be obliterated for generations from places that had been their own. Some seeds, acting on ancient orders, were habituated to long waits in the soil. Many of them passed through wet periods without germinating. This was the ultimate mystery, surely, an inexplicable restraint against movement when every fact cried out for action.

There were other mysteries. Some grasses sprouted on the plains only periodically, at far-spaced intervals, without demonstrating any reason for their abrupt appearances and disappearances. One patch of ground which this year had seen a swarming growth of an annual of one type would not host that grass for the next two or three years. Instead, in its place, two other species of grass might rise. Such rhythms were connected, somehow, but through channels that were inaccessible to ordinary analysis.

The moon had risen. The leopard's yellow eyes gleamed. The baboon's exploring eyes focused on the leopard's uprising paw, saw the spring-coiled body unleashing in slow motion as reflexive messages flashed through his brain in a fraction of a second, pumping strength into every muscle. He was no coward in the ways of the grazing animals of the plains, who might run blindly when threatened or who jumped vertically in useless, stationary stotting, or pronking. He was herbivorous, but also fitted to fight ferociously. The lifting of the leopard's paw triggered the first visible reflex, the opening of the baboon's mouth, which revealed lethal yellow fangs. The second reflex was the raising of both the baboon's hands, ripping claws fully extended and moving to parry the leopard's blow.

With all the supercharged juices of survival exploding inside his body, the baboon's hair bristled so that he became almost twice his real size in an intuitive effort to deceive the leopard into assuming real danger. At the same time, the first strong gust of night wind reached into his fur, rubbing against his back so that he looked even larger.

The night wind, which came almost every evening, was a benign wind that blew impartially for all living things, but some knew how to extract advantage from it. Wide-ranging moths, creeping out of secret hiding places in the crumbling earth, were carried miles away downwind to new refuge places for the morrow. Night-flying birds, particularly large owls, used the wind to travel hundreds of miles, crossing the plains in soundless flight and hardly ever needing to flap their wings.

But it was the grasses which benefited most from the wind. It provided relief from the heat of the day, blowing as the temperature fell, blowing as ten thousand square miles of condensation collected between every particle of soil, every grain of sand, on every leaf and stem, every twig and branch. The moisture ran down eventually into the soil and there the annuals absorbed it immediately through cotton-wool-like masses of fibrous roots.

The leopard's paw, now in motion under the full impetus of his expanding body, would strike the baboon between ear and neck, a lethal stroke which should break the spinal column while the agile left paw pinned him. But the baboon had turned his face toward the oncoming paw instead of trying to avoid it. The claw-filled paw was now on an irreversible collision course with a face of fangs. Milliseconds before the impact the baboon threw back his head, not to avoid the blow but to ensure that his fangs would penetrate deep into the weapon seeking to kill him.

When the collision came, the two large incisors on either side of his mouth sank into the leopard's paw. They slipped between the outstretched claws and buried themselves in the pad behind. But the claws, outstretched, drove deeply into the muzzle and lip of the baboon. The force of the blow knocked him backward so that he had no immediate chance of recovery.

But the impact of his teeth saved him, because the surprised leopard felt the pain immediately. He knew the strength and determination of baboons, but this time he had assumed total surprise and was not expecting retaliation. The launching of his left paw was delayed for another second while his brain absorbed the messages of pain from his right paw. In that moment, the baboon staggered backward and got his front paws to the ground behind him. He half flipped, slipped, slid, and nearly rolled over the edge of the rock, but got a grip and was running before the blood began to pour from his face.

One image remained supreme in his brain: the dim and

sheltering trees in the center of the island. No such over-riding image occupied the leopard. Unlike the baboon he had not been under intense stress at the time of the attack. His recovery, therefore, was slower. By the time he had reacted to the savaging of his paw, to the failure of his belated swipe with his left paw, which had pounded on bare rock, the baboon was gone.

No baboon could outrun a leopard, who in short bursts moved as fast as a cheetah. But the night and the bulking rock made it impossible for the leopard to charge freely. He paused, listening to the scrabble of feet, heard the rip and tear of claws on bark, then saw the baboon's black body hurling itself among the highest, thinnest branches of a tree in the middle of the island. The baboon stopped when his weight bent the branches beneath him. He looked down, dripping blood at the feet of the leopard.

The leopard realized the impasse. He could climb the tree. But he could not climb that high or maintain himself on such thin branches. He knew nothing would dislodge the baboon. His only chance was to outwait the treed animal. His paw hurt and blood pumped from it. He licked it. The wind grew keener, the plains became smoky gray and faintly luminescent as the moon rose. A rumble of thunder, or perhaps it was the combined voice of lions, came out of the dark, and the wind whispered secrets to the island around him.

III

When the vultures lofted from the island's trees, usually soon before midmorning when the night air had warmed, their vision of the plains expanded a score of miles in every direction. They could see solitary, flat-topped trees and the rocky bulge of small islands rising from the golden landscape. To the north, the well-treed water hole and its protective islands were visible. To the south, the faint mark of blue hills showed the termination of the plains in foothill country where lay the extinct volcanoes which had created the plains. To the east, the horizon was empty except for the odd tree. To the west, the rounded structure of another large island arched from the flat earth.

But for any vulture who drifted farther north, the landscape gradually changed and became more diverse as the bare plains gave way to a scattering of islands, to dried watercourses flanked by thickening growth, to bands of acacias marching in linear formation across the flatness, to occasional greening where moisture lingered. The plains became less flat, undulating gracefully in places, and revealed small groups of gazelles, a solitary rhino, a pair of cheetahs slumped in sleep, a pride of lions sprawled among the trees, giraffes camouflaged in groves. Eventually, the plains turned into open, well-wooded savanna country in which neither grass nor tree held total sway. There, hills and ridges, odd swamps, long lines of winding pools and rivers, some of them temporarily stagnant, created a variegated landscape that eventually terminated along the shores of a vast lake stretching beyond the sight of the sharpest vulture eye.

The northern flight of any vulture revealed the contrast between the emptiness of the plains and the life in the woodlands. But it told nothing of the many stirrings and movements that were soon to transform the woodlands and the distant island. These hints of change signaled the beginnings of a unique season into which would be compressed the essence of search, flight, procreation, birth, death, chaos. Yet, at this moment, nothing about the north suggested such drama.

Its pleasant mixture of well-spaced trees, its grass, thick and dry, its rolling hills and many watercourses indicated superficial differences from the empty south but gave no sign that its relationship to the south was symbiotic. Neither could exist satisfactorily without the other. The two were matched like gears. The north held shelter and most of the options for survival in drought. It protected a huge reservoir of animals, two to three million antelopes and horses. It was the practical place, pragmatic, where systems of life were explicable, sensible.

By contrast, the plains represented the romantic and the irrational, the illusion of life. They contained a dream of great deeds where the weak sometimes triumphed over the strong, where abundance might flourish, where sex and birth and death could coexist on a mammoth scale, where great gatherings of animals might parade in seeming contentment and camaraderie.

In the meantime, though, the north remained the place of reason, where rains had sponsored the longest and coarsest grasses which grew six feet high in the best of years. It was this growth, dried into hay fodder, that provided the food to sustain millions during the drought. The north supplied nearly all the dry-season food; the twigs and shoots and shrubs on which rhinoceros, impala, eland, bushbuck, topi, baboon, and others browsed when the grass disappeared or became too desiccated to eat.

It furnished shelter for the secretive elands, the largest antelopes, and also for the smallest antelopes, the duikers and dik-diks, the klipspringers, the sunis, the steinboks and oribis. It provided trees for the long-reaching giraffes and elephants, fruits and flowers and seeds for a dancing multitude of birds: laughing doves and warblers, cisticolas and prinias, sunbirds and widow birds, woodpeckers and scimitar-bills, kingfishers and mousebirds, trogons and camaropteras. It was a refuge for monkeys and vampire bats, for ant bears and bush pigs, for aardwolves and otters, for shrews and hedgehogs, for porcupines and cane rats, for spring hares and dormice, for giant rats and palm civets. Its flowers nurtured a constellation of nectar-sucking insects, and its animals hosted equivalent armies of blood-sucking flies. It seemed to be self-contained, but it lacked a commodity that the plains possessed to complete the symbiosis.

The grasses that grew on the plains around the island were different from those in the north. Sweeter, shorter, easier to eat, they did not shoot up, lank and unpalatable, when rain came. Instead, they could be grazed again and again, each time recovering swiftly from the clipping of their stems and leaves. They gave sustenance first to the high-cropping zebras, then to the wildebeests, then to the close-cropping gazelles. Finally, they fed locusts and other insect-grazers, termites and beetles.

Now, the island waited. The injured baboon nursed his mangled face and cautiously browsed residual seeds from the topmost branches of his refuge tree. The leopard slept; the hyraxes fed, and the sense of anticipation increased throughout grassland and woodland. The northern animals moved into new associations, either dividing themselves into smaller groups or amalgamating into larger corporations. The various races of zebras—and there were many, separated from each other by tiny differences in behavior—began tentative migrations, straggles of movement in the direction that their later, formalized migrations would take. The races had mixed during the drought, more or less, but in their separations some drifted south toward the plains while others went north toward hilly country.

The wildebeests, those most mournful of all antelopes, heavy-shouldered, with downcast heads and drooping horns, their black manes and white beards making them look more like crude horses than their daintier close relatives, the gazelles, had gone through the drought in their usual herds of a dozen to several hundred. Now, they were forming into even larger groups as they continued their ever-lasting circling in search of food. They followed such well-marked routes that an old lion, who had spent much of the drought time stationed along the territory that divided the woodlands from the plains, could place himself in their path and be sure they would appear, if not today, then tomorrow or the next day. An old hyena, her den located in nearby territory, led a group of hyenas into the night woodlands and attacked one of these growing wildebeest herds. She, like the lion, awaited the movement of these animals, anticipated the time when they would come to the plains and she and her kind would feast.

The small gazelles, with their graceful, slender legs, their relatively tiny, single-striped bodies and short, elegant horns, were diminutives of the larger gazelles already out on the plains, but were just as much antelope as were the wildebeests. Now, they bunched themselves closer to the zone of division that separated plains and woodlands in readiness for the movement that would take them, more than one million bodies, to the magic south.

Each of these groups of animals remained bound by the laws of the season which held them to the long-grass country. But within their social orders, changes were occurring to prepare them for their time on the plains. The bucks and bulls were daily becoming more insistent on maintaining their rights over their females. The first of the serious fighting had already begun among the zebras, who would take with them the most cohesive family groups of all the migrants. The buck gazelles who had held territory throughout the drought were made indecisive by the need to move and by the equally insistent demand to possess territory.

The new restlessness of the grazing animals affected the meat eaters. The lions, in particular, were bound to their territories in the long-grass country. They must remain there when the grazing animals left. They were anxious and nervous, and fights between rival prides exacerbated a dilemma only dimly perceived.

Even at this moment the interconnection of northern grasslands and southern plains was an unbroken bond, the ocean of grass and the distant island imprinted on a multitude of memories.

As the animals waited, one of the few remaining sources of water dried to a series of long, old-blood-colored pools. The quiescent river halted abruptly at sandbanks, its dried course disappearing under the dappled shade of thriving fig trees. Six motionless geese floated in one of its pools. At the water's edge sandpipers rippled the shallows and flew like large insects, intense in their movements. Small bee eaters dashed low across the water. Lovebirds shrieked, peach-colored faces caught in the sun. Fiscal shrikes streamed long tails from tree to tree. Baboons and impalas, nervous as always, came to the pools to drink. The somnolent buffaloes, who cared about nothing, switched dry-season flies from their flanks. Giraffes and warthogs appeared silently. Each, in his circumspect manner, drank cautiously.

The pools looked empty until one heaved and a hole appeared in it. Air whistled out of the hole, and the nostril of a hippopotamus closed and disappeared. In another, a solitary green eye opened, a chill relic from the age when reptiles dominated the earth. The days passed in an endless succession of white-splashed blue skies, and thousands came to the pools. One by one, and in small groups, the drinkers arrived in a steady stream. The ground was pulverized into fine deep dust, overprinted with footmarks. Piles of dung glistened in the sun. Drag marks erased footprints where crocodiles had moved. A baboon barked alarm. Impalas fled.

Tracks cut out from a pool into the wooded grasslands on one side and into scattered trees on the other. They wound along the edges of dried watercourses and came together in sudden bold collisions which fanned out into highways leading toward the grasslands. These were not yet in full use.

Vignettes of the woodlands suggested an eternity of life there, a visual counterpoint to the island, brooding alone to the south. A lark hung in midair, beak full of the purple manglings of insects, and a large hawk hung behind it for a second before sweeping away among the trees at high speed. A green-backed bird, eyes striped, underparts yellow, triangular wings black-barred, fired away from a dried stem, captured an insect, rocketed vertically to catch another victim, then returned to beat it to death on a twig. The hammering done, he took off again, a sunbird facing into the eternal wind. Throughout the wooded grasslands the green of the tree foliage soothed all the senses. The sun passing through leaves took on and kept cool green tints. Among the yellow-fever trees—one of the acacias—the yellow-lime trunks created color groves where the sunlight, already dappled through the green foliage above, seemed turned into another form of light as it bounced back and forth among the delicately colored trunks.

The glowing light created an eerie, expectant peace, broken only by the soft rustling of seed eaters in the canopy of trees. Long-bodied touracos hauled clumsy tails across clearings, screeching at mongooses beneath them. Would it rain, would it not rain? The cries questioned the waiting air and reproached each cloudless day.

Young zebra stallions broke the tension by fighting and by abducting mares. The wildebeests sharpened sexual differences. Though no rain had fallen, bulls withdrew from mixed groups and joined all-male herds. They fought often to establish territories, scraping the ground vigorously with their feet or rubbing their foreheads into the dry earth. Any cow wildebeest passing through such male herds would be importuned a score of times in a day by anxious bulls. The establishment of territory, in the face of the certainty that all the wildebeests must soon move, seemed to contradict the purpose of it.

Before they felt the oncoming season, the gazelles had been divided into male and female herds with some of the males seeking to establish territory in the woodlands. But as anticipation grew the gazelles changed. Males joined females, territories were abrogated or relocated, and these small antelopes began gathering in spasmodic movements toward the south, often in herds numbering hundreds. Their fluid mixture of sexes and territories contrasted with the elands, who remained in stable herds, traveling widely and secretively, and who would remain in these herds when the challenge came to move south.

The overwhelming numbers of grazing animals—more than a million gazelles, nearly half-a-million wildebeests, two hundred thousand zebras, and several hundred thousand other antelopes—contrasted strangely with the relatively small numbers of the meat eaters. Only one thousand lions lived in all the woodland prides. Fewer than five hundred cheetahs, about five thousand hyenas, a thousand dogs, four thousand jackals, and a few leopards lived among this feast of grazing-animal flesh. But the numbers of grazers were not measurably influenced by the meat eaters. Instead, they were controlled by creatures who killed with invisible stealth.

The anxiety which preceded the new season was felt most acutely by the old animals. In the woodland herds they faced the difficulty of trying to join bachelor herds which might not accept them. If they were allowed to join, they would be among younger animals fighting upward in a power hierarchy. Now, lone zebras fell out of family herds, their strength gone because their teeth had worn out eating the abrasive dry grass. They were starving and the lions picked them off. The buffaloes, great community animals, would not accept old bulls in their midst—even though no migration faced them—and the old banded in separate herds to collectively resist the prowling lions.

Both young and old died in this time of change. The obvious deaths came in the struggle for power, territory, food. But by far the most common cause was the work of the creepers and the crawlers. These creatures—organisms —infested almost every part of every body. Livers, kidneys, lungs, nostrils, eyes, stomachs, intestines, bones were all vulnerable. An old animal, still relatively strong, still anxious to live, found his vital forces diminished enough to give that final parasite that final foothold which would kill him. Here was a stealthy world of silence. Instead of the noisy, yelping hyenas and the dramatic leaps of leopards, there was only the hidden sucking, biting, burrowing, poisoning, consuming work of the parasites. They had such power that entire nations of animals could be wiped out by

40

them, thousands killed in quick successes of a single kind of parasite.

The wildebeests trudged along, maned heads stolidly down, lungworms working inside their bodies. Some resisted the worms' work, others sickened and stood helplessly while their fellows passed on. Ticks teemed everywhere, vectors of countless blood parasites which moved from animal to animal by unplotted roadways. Tapeworms worked through the livers of impalas and reduced their soaring leaps to feeble stumbles, bringing the graceful animals within reach of hyena and leopard. Giraffes hosted nematodes in their livers, parasites in their bloodstreams, and a swarming of skin parasites which sent clinging birds— the oxpeckers—exploring all over their bodies in search of ticks, eggs, larvae.

The island and the plains waited for the rains to begin and for the visitors to arrive. What stood between the antelope and horse and their journey to the land of promise? Gedulstia flies laid eggs in antelope nostrils, where they hatched and the maggots worked up into antelope brains, causing meningitis. Like the stricken wildebeests, the antelopes stood helplessly and turned their heads repeatedly from side to side as their comrades walked on, leaving them to die.

Nothing escaped the parasites. But the animals had lived so long in woodlands and plains that they had learned to function even with heavy infestations. Ticks smothered buffalo and eland, giraffe and antelope. Lice were everywhere. The moment an animal sickened slightly, their parasites flourished responsively so that they, rather than the original disease, might kill the animal. The topi and the kongoni, the wildebeest and the gazelle all harbored the larvae of many different creatures crawling through various parts of their bodies. Fly maggots worked into the frontal sinuses of antelope heads. Nothing escaped, nothing. Stomach flukes and liver flukes, lungworms and tapeworms, they penetrated muscles, lungs, brains, livers, kidneys, intestines, feet, ears, nostrils. They bored inside the sacrum bones of some of the antelopes. Many animals harbored simultaneously the egg, the larva, and the adult stage of the parasite.

The animals moved on through the crackling woodlands, heads down, tails twitching. Tantalizing dark clouds passed and repassed, but nowhere was there a sign of rain. The sound of flies' wings was everywhere. The warble flies attacked mostly at the rear. They laid eggs around the flanks and legs. Botflies attacked at the front. The strategy of each fly was different, but the objective was the same: to insert the fly's progeny into the body of the victim. The warble eggs clustered on hairs, hatched, and the maggots bored down through the skin. Once inside, they could move at will, although they had an objective, the gullet. To get out of the body they did not simply leave by the nearest exit, which would have been through the neck. Instead, they made complicated migrations to reach places just beneath the skin of the animal's back. There, they finally matured, cut through the skin and dropped to the ground where they bored holes, pupated, and so were transformed into flies again, ready to reinfect their hosts.

The botfly maggots did not hatch spontaneously but waited to be moistened by the animal victim, who licked the eggs or rubbed them. The eggs, hatched with larvae, were swallowed and attached themselves to stomach walls where they became blood feeders. No migrations or com-

42

plicated journeys for them, just simple ejection through the intestines with the excrement, and so back to the ground. There, they followed the warble flies into pupation and repetition of the cycle.

Wherever the host animals moved the parasites followed on wings, or traveled inside them, or fastened themselves to fur and hide as the hosts passed through shrub and grass. Parasite competed with parasite for a share of the victim. Honey badgers, dangerously tough and ill-tempered, with overpowering appetites for honey, had to share these sweet meals with tachinid flies who were now especially busy because those acacia trees which anticipated the rain by flowering early were swarming with bees making honey throughout the woodlands.

During these busy days the tachinids hovered at hive entrances, sliding horizontally back and forth in movements that looked aimless. But for the honey badger about to climb the tree, they meant the hive was healthy, therefore it had honey. Each female fly touched a bee as she darted forward, the movement so precise that the bees were not distracted. Tachinid larvae, liveborn, were stuck to the bees with dabs of adhesive.

The maggots worked into the abdomens of the bees through division lines between the segments of their bodies and ate each bee alive, ingesting the abdominal fluids until the parasite almost filled the bee's body. This did not seem to disturb the bee, and the parasite left the comsumption of the most vital organs until the end. Abruptly, the bee fell, buzzed, died. Perhaps the larvae released poison which gave them precise control over their victims. They emerged, buried themselves, got ready to pupate in the inevitable cycle of the parasites' lives.

The badgers seemed to do the worst damage. They tore their way into hives and scooped out all the honey, destroying the hive. But they had little effect on the sum of all the bees in the woodlands. Only about one thousand badgers worked the woodlands, and each pirated fewer than a score of hives a year. It was the tachinids who killed countless millions of bees, thus arbitrating the numbers of hives and rationing the badgers' appetite for honey.

In a season of imminent change the life cycles of many of the parasites had to harmonize with the movements of the animals. It was pointless for the larvae of some flies to drop from their host animals far out on the plains where they would have to pupate in hostile country. Most of the parasitic flies depended on the shelter of the woodlands. Their work, therefore, now had special urgency as the animals prepared to move.

Others, however, traveled permanently with their host animals. Nearly all the animals had fleas and so were susceptible to plague, perhaps the most prevalent of all the diseases. But most animals could withstand this infection. Only when other, massive infections or parasites afflicted the creatures might plague become a killer. Then the fleas, carried a hundred miles into the plains and deserting a corpse already being rent by scavengers, might have difficulty in making a quick transference to a new host animal. The fleas, of course, were infected with plague. Germs clustered so thickly in their systems that they blocked the fleas' digestive tracts. The fleas, already ravenous because of their inability to digest, drank the blood of their new hosts—hyena, lion, jackal—but were unable to get it into their systems. They starved, injecting plague bacilli in the process.

The work of the parasites sent regular, rhythmic shock waves across the plains and woodlands. They were one of the oldest forms of life and skilled performers, functioning with subtlety and success. Their hidden power helped to maintain the impenetrable mystery of life here. Elephants bulking through the dry woodlands and smashing down a hundred trees a day in search of foliage might not fall to lions, but they could not repel the parasites. Some of these, maggots bunched together in masses as big as a starling, ate away at the elephants' tissues close to their hearts. Others, particularly the warble flies, sent maggots into the flanks, shoulders, thighs, backsides. One parasite even managed to bore into the iron-hard tissue of the elephants' feet. The botflies could put their maggots deep inside each elephant, under the skin of the ears, or high inside the trunk, a supreme aggravation. The parasites respected nothing. Their ageless task was to monitor all life. Whenever their host animals became too numerous, were weakened by other diseases, or otherwise disobeyed the orders of a balanced life, they surged forward to restore equability. And so there were never too many lions, too many wildebeests, too few gazelles. The balance appeared precise.

Woodlands and plains thus contained a perfect deception where the eye saw nothing but lies. The leopard's bark sent terrified gazelles spinning away in the night, and that appeared to make the leopard important. The rumble of lions agitated the grazing animals, and the sound, being powerful, appeared to signify powerful animals. But such externals did not resemble the internal order, the invisible rules and unfathomable imperatives.

Some parasites were able to work at least partially independent of the season because their victims would not be migrating. The old lion lay in the shade of trees, his bloodstream teeming with tiny protozoans, members of a great underkingdom of single-celled animals with elongated, slender bodies who were able to whip themselves through blood corpuscles by lashing their tails. These creatures, the trypanosomes, penetrated the bodies of mammals and birds, of reptiles and fish. The animals had made various adjustments to the trypanosomes so that the omnipresent parasites usually left their hosts unharmed. They were carried by more than a score of different flies, though they were most partial to the tsetse.

The tsetse was a perfect instrument for the parasites. A drab, yet oddly beautiful fly verging on the sinister, he had yellowish tinges on his body, a touch of red in compound eyes, a proboscis pointing forward, and wings folded back over the body like a pair of scissors. He was a patient blood seeker. He would wait long hours for blood in camouflaged places against the trunks of trees. Each fly had sensors able to measure heat and humidity, to receive sounds and gauge the intensity of light. Their sensors rigidly controlled their travels, preventing them from venturing far from cover. They could never leave the encompassing embrace of the trees. A treeless clearing only a hundred feet wide stopped their movements completely. But the sensors also gave them an uncanny capacity to find precisely the animal they needed to get blood.

Their activity matched the growing restlessness of the animals. Once the migration began, the bulk of their blood meals would be gone. From gazelles and wildebeests they would be forced to switch to more elusive prey; impala and topi, lion and leopard, warthog and hyena. More important, they thrived in drought. Once the rains began their work would be slowed or stopped, themselves become partially comatose until the drought returned, and with it, the millions of prey animals who brought them blood.

So now, they hunted quickly. They slipped away from watching positions in trees and flew low and direct for their victims. Their attacks were always at grass level so that they came up underneath their victims without giving warning of their arrival. Immediately, they inserted probes through the thin skin of the belly.

The victim usually did not feel the fly's proboscis drive in. This was no random quest for blood. The probe moved around, curved, moved to one side until it located a suitable capillary. Then, a sawlike appendage concealed inside the probe quickly ripped the capillary and began the flow of blood. During the incision period, tsetse saliva coursed freely into the wound as a lubricant and anticoagulant. The blood, once tapped, would continue to flow.

With the capillary cut, the gush of blood filled the incision. The pumping muscles of the fly went into action, ingesting the blood. So rapid was this action that within minutes the fly was blown up tight with blood, body colored dark red by the quantity of liquid packed inside.

The tsetses withdrew their probes and remained motionless for long moments on the trembling bodies of their hosts. They were unable to fly immediately because of their burdens of blood. When they launched themselves, they often crashed, the blood load too great to transport back into the shelter of the trees. There, supremely vulnerable to any ground blood hunter, they had to remain still while their blood-processing systems were galvanized. The blood moved into the crop, then into the midgut, the anticoagulant preventing it from slowing its passage. This enabled the flies to quickly extract the water from the blood, pass it through their cells, and expel it from their anuses in droplets of waste matter. The flies could absorb and process, or dismiss, one quarter of the entire weight of the blood meal within fifteen minutes of feeding. This efficiency gave many of them a second chance at survival as their enemies came to them in the grass.

Because the tsetses found their food by sight, impelled to hunt by a flicker of fur among the long grasses, they were led on by the moving animals. They concentrated along the game trails that led to water holes, frequently losing their victims when the animals spread out to begin grazing or moved during the night. Yet at least half the time the flies found themselves at water holes, where they collected in thousands. But with the water holes drying and the river pools shrinking, the tsetses became progressively more savage in their attacks. The first antelope to appear in the morning would be smothered with hungry flies, sometimes landing so thickly that the animal would be driven to flight as ten thousand probes, ten thousand saws, sought to cut at his body.

Hunger grew among the tsetses and they followed the trails of the animals from the water holes, spreading among the woodlands to saturate every area where there was blood. But the tsetses were themselves pursued with a similar determination. Robber flies followed them the way cheetahs hunted. The robber fly selected its victim, following every move of the zigzagging tsetse until it alighted on its host animal. A second later the robber fly arrived, enclosed the tsetse in its embrace, and flew off with it.

More deliberate in their predation were the hunting wasps who moved slowly among the trees in search of resting tsetses, inspecting trunks and the undersides of limbs with great care as they sought to discriminate between the background and the camouflaged fly resting in it. Spiders spun webs strong enough to hold tsetses. Blood-loving birds, the drongoes, ate them. Flycatchers were aloft in constant watch for tsetses moving from tree to tree.

The tsetses, carriers of rich cargoes of blood, were pleading to be victimized themselves. Whenever they rested in trees, digesting with their heads up, or waiting and watching with their heads down, small spiders stalked them. The tsetses' eyes ranged widely but the stealthy spiders inched to within jumping distance, then leaped and enveloped their victims in silken bonds. In all the five thousand square miles of the woodlands, eight million tsetse flies were killed every day.

Despite the serene landscape of golden-brown grasses winding through graceful trees and skirting green-brown pools, of eternally blue skies holding the hanging marks of vultures, of blotched lines of grazing buffaloes on the horizon, the trypanosomes and the other parasites were the truth of it all, as firmly connected with the waiting island as the intervening grasses. The tsetses reached out for the island. The parasites swam in the blood of the leopard and of the baboon. Every hyrax carried them, and so did the striped hyena. They would go to the plains in the blood of the grazers.

The tsetses displayed themselves in this African landscape with a clarity that few other creatures provided. Their lives indicated the complexity of all life, and this despite the fact that they were lowly flies. While the zebra stallions consolidated their families, the gazelles changed their herds, and the wildebeests formed into bigger groups, the tsetses were mating.

The urgency of their mating suggested the imminence of

seasonal change. The females could mate only three days after they had pupated in the earth and emerged as winged insects. On some days millions of males would be seeking females, following them in wild chases, seizing them in midflight, the competition for females so fierce that three or four males might simultaneously be grasping the female fly. Clusters of flies tumbled end over end and fell to the earth, until one male triumphed and mated.

The act of insemination, evolved in an intricate patience of time, could be described but not explained, any more than the leopard's presence on the island was explicable, or the baboon's plight was sensible. Life had thrived so long in this landscape that its methods were arranged into an incomprehensible mosaic. The mating tsetses remained locked together for much of a day, prolonged perhaps because the female would be fertile afterward for the rest of her life. She stored the male's sperm in two sacs connected to the top of her uterus by passages down which she could direct sperm at will. Blood was the food, an inexhaustible supply of it, and the flies thrived into expanding populations. But their great and growing success gave their own enemies chances to kill large numbers of them. So the tsetses had to stretch ingenuity to protect their children, both egg and larva, from the many who hunted them. Instead of laying eggs like conventional flies, the female brought both eggs and young larvae to maturity inside her body. Then, as though imitating higher forms of life, copying the mammals on whom she preyed, the female gave birth to living young.

Before birth, one egg hatched inside the body of each female. The larva used an egg tooth to split open its shell, as though, unexpectedly, it was imitating birds. It changed its

50

skin immediately, dropped its egg tooth, exactly as young birds did. But then it fastened itself to a teat at the bottom of the uterus, like a marsupial youngster. Twin milk glands provided nourishment so rich that the entire growth of the larva was compressed into five days, a speed of development already suggested by the fact that some females were only sixteen days old when they began giving birth.

The moment of birth was critical. The tsetses had to find correct places to drop their young. The flies flew so secretively to their maternity grounds that few predators ever saw them. Their sensors located exactly the right spots to land, the ground beneath shaded and soft. Then, contracting their abdomens, shaking and quivering, they finally bore their young. Gross creatures, these, out of proportion to the size of the fly, but energetic, and with steady contractions they wriggled out of sight into the earth.

The perfection of tsetse development accorded homage to the complex adjustments of long evolution. But it was still not perfect. Sometimes a female, nearly ready to give birth, could not control her greed for blood and took one last great meal which so distended her body that it trapped the young and made immediate birth impossible. The mother could neither give birth nor excrete the waste matter from her blood meal. Helpless, she flew. Helpless, she waited while the larva, unable to be born, pupated inside her. Helpless, she lingered fourteen days before dying with neither of the processes completed.

The tsetses might well make such careful preparation for protecting their young. Almost every inch of soil was searched daily by all the ground birds—the francolins, the guinea fowls, the bulbuls—and by the mice, the elephant shrews, and the ants. At least twenty different kinds of wasps and bees, ants and flies sought to reach the subterranean tsetses, but locating the pupae was hard. Only one in thousands was ever found by hunters. Some tiny parasites, wasplike in appearance and only one millimeter long, explored the soil to find pupae. If they were lucky, they bored a hole in the pupa and deposited thirty to forty eggs. But they were poor burrowers; they could reach down an inch and a half through dried leaves, but only about half an inch in dry loose soil. Another wasplike parasite, who was hunting quickly before rain ruined its work, could introduce only one youngster into each tsetse pupa. These females were so specialized in their search for the tsetses that they had sacrificed their capacity to fly so they could burrow better.

The end of the drought was anticipated by multitudes of animals as food grew more scarce everywhere. On the island, the baboon sat trapped in his tree. He had eaten every available twig, every available seed and piece of dried fruit. Among his own troop of creatures, he usually roused himself late, sometimes after the sun had appeared, but now he stirred before dawn when the leopard's branch was empty. Then, feverish, he dropped to the ground and rushed among the long-dead grasses on the western side of the island, licking the remnants of the night dew from their stems. Each time, though, the leopard had appeared. Once, he peered over a rock; another time, he looked down on him from a nearby tree. The baboon was frozen in controlled panic, but the leopard had not moved. The hyraxes, caught in a trap of double jeopardy—the baboon their enemy as well—kept the days ringing with alarm whistles. Each hour passed in the dread of the unexpected, as if the invisible conflicts of the woodlands were reaching out and touching the island. Life moved forward, one second at a time, and the leopard slept, and he did not sleep.

A final force gathered itself for a descent on the island. It traveled with a group of elephants making a long migration from the far north where the drought had endured two years. They passed through baobab country, open dry plains studded with short-branched trees which stored old rains in the fibrous interiors of their monstrously oversized trunks. The elephants ripped off the branches, stuffed them into grimacing mouths, pushed over entire trees. Then they ate them, trunks and all, masticating the wood into pulp, spitting out the fiber in large yellowish balls.

With the elephants came anthrax, a patient traveler unaffected by shortages of water which drove the thirsty elephants to plow long furrows in the dusty ground. The anthrax had a perfect host, a herd too big for its own security, a self-contained world of victims in which the disease could go from animal to animal. From a hundred animals at the beginning of the migration, the elephant herd had declined to fifty. As the trembling victims fell, belching and hemorrhaging, bulls let out trumpeting screams while cows and youngsters milled about nervously. Each time an elephant died, the others took final and reluctant leave of the corpse. When they reached the northern grassy woodlands, eight animals were left.

In its slow, southern movement, the anthrax provided poison bait for woodland animals. From a dead young elephant, it passed briefly into the old lion, but did not infect him. From his teeth, it went into the body of an oryx. The remains of the oryx infected three lionesses. The strain of anthrax was deadly, or had managed to transmute itself in the exchange, because the lionesses did not have time to leave the kill before they were stricken. Trembling, unable to move, they lay beside the oryx rib cage and skull and died one by one.

The anthrax passed into the soil and touched the dried grasses impartially. It cut down the tiny dik-dik and the lumbering eland. It met and melded with migrant strains which had lurked, quiescent, in the woodlands. Like the other woodlands diseases, it was stimulated by the length of the drought and struck into a community of animals already touched by shortages of fodder and water.

But the woodlands were wide, and the animals remained well spaced. Herds were scattered and families destroyed, but the anthrax could not get among enough animals in one place for its effects to become a catastrophe. Its problem was to keep its strain alive until the rains started, until the migrants began moving and concentrating. Then it would be able to travel and to proliferate at immeasurable rates. It could reach the island. Perhaps the sick elephants would be responsible, in the end, for saving the baboon from the teeth of the leopard. By this labyrinthine highway of disease, they might infect the big cat but spare the baboon. Or the baboon might die and the leopard be spared. Or both might die.

The island waited. The baboon waited. The leopard waited. The plains waited. The skies waited. No sign of rain appeared anywhere.

IV

The old lion, a somnolent watcher of the endless sunny days, groaned uncomfortably in his semisleep in the shade along the banks of the half-dried river. Age lay upon him painfully and was measurable by the increased weight of his paws, each one striking the ground clumsily and noisily when he walked. His head was heavier and he carried it lower than he had the year before. His face had changed, becoming longer and thinner, and expressed a suffering that was neither pain nor fear. Sometimes, when he got up, parts of his pelt lay scattered around him as if the pieces had been clawed off in a fight. Parasites swarmed in his skin. He could feel ticks gathered in his groin. He had quit scratching

55

his patchy mane as the number of parasites working away in that refuge grew far beyond his capacity to exterminate them.

He was far from the island on the plains, but as a nomad, a solitary wanderer in every part of the plains and woodlands, he had been there scores of times. He knew every part of the plains in all its seasons. In his wanderings, during the good times, it usually took him fifty-five days to circumnavigate the plains and return to the woodlands. Now, however, plains hunting was near impossible in the extended drought, and his living was precarious enough in the dry woodlands where his natural victims, the small gazelle, the wildebeest, the eland, the topi, the oryx and bushbuck, and countless other smaller creatures were either spread out widely or were sheltered by the abundant vegetation remaining there.

He, too, like all the other animals, awaited the beginning of the rains. They would consolidate his victims into marching masses which could be attacked profitably. The rains would bring the animals out of the woodlands and into vulnerable positions on the plains where, preoccupied with calving and rutting and grazing the sweet grass growing there, they would not see him with such urgent regard. With the rains he would become a true hunting lion again, able to feel the power of his presence among all the lesser animals around him. But for the moment, he had little to do but sleep; heavy, hot sleep that somehow did not rest him. In such sleep, becoming more common each season now that he had grown old, he often fell into that state which was as close to true sleep as any cat can ever achieve. In this dreamlike mood, he seemed to go back in time to lie in solitary magnificence at the edge of the plains, his mane as thick and black and glossy as the peak of youth and health could make it.

His role as pride leader had been ambiguous. Undoubtedly it was based on his size and power, qualities which had gained him ascendancy over other lions and lionesses, but he had spent little of his time physically defending any territory. Rather, his success had been measured by his choice of territory, if indeed it was choice, and by his attention to the task of marking its boundaries with scrapes of his claws and with sprays of urine which warned intruding lions that they were entering alien country. It was measured by the power and quality of his roaring, which communicated his determination to hold the territory sacrosanct.

And yet, despite these manifestations of authority, he rarely led the pride anywhere. Rather, he trailed his hunting lionesses, if he followed them at all, in what seemed to be a subservient position, but which was shown to be protective when the cubs of the pride flanked him behind the lionesses. He appeared to be the leader, but he could be intimidated, even driven away, by a group of his own lionesses suddenly unwilling to accept him. He and his companion seemed to be the stabilizing influence on the pride, its protectors and fighters against the dangerous outside world, but his companion might unaccountably disappear, to be replaced by another male perhaps hours or many days later. He appeared to be the defender of the territory and its guarantor, but it was his lionesses who were the more vicious in their refusal to allow new lionesses to enter the pride. He seemed to be intolerant and completely selfish, and yet he would accept into the pride an intruding lion, maybe even two or three of them, if he sensed a lack of danger in their voices, scents, or behavior.

57

As the leader, he could afford to be indifferent to the manifold problems of the hunt. Whenever he had been hungry, he had merely taken the kill away from the females. This appeared to be his right and his reward for correctly establishing a good territory along the division line that lay between the woodlands and the great plains. But while he gorged himself on the carcass and growled furiously at the surrounding lionesses, he would allow some or all of the pride's cubs to feed with him. It was the lionesses who fed the more selfishly. Each animal in the group had his role, precisely delineated in the long stretches of time that had brought the lions to their place of dominance on the plains and in the woodlands.

His security as a pride lion had made him conservative, unwilling to move from place to place with the freedom many other lions displayed. The success of his pride had kept him bound to the territory which had made that success possible, and so he remained a fixed force in the fluid and flexible movement of lions around him. When the antelopes moved to the east in response to unseasonable rains, he had remained in his territory while other members of his pride drifted away. When the animals left the plains, or migrated toward them, he had stayed in the one place and watched his pride swell as others joined it to prey on the avalanche of animals moving through his country. He had been a leader, but the true nature of his authority was obscured by the complexity of lion society.

He remembered sitting in the sun, much as he was doing now, magisterial in his dignity. Young lions always played near him; some of them would be nearly three years old and their playing was rough. The victors in these play-fights were the lions he would have to fight later, driving them away. They could never be reconciled with his authority. They would not be powerful enough to fight him successfully, but they would be too strong to be tolerated as a sexual presence among his lionesses.

In the midst of his sleep, he scented hyenas and lifted his shaggy head, trying to sense whether their proximity meant they had fed, or were hunting, or just intent on making trouble for him. Once before, he had been harassed for half a day by a large group of hyenas after he had been injured in a fight with an older lion. He had been furious, and in one quick and savage swipe, he had broken the back of one of the animals. Its screams echoed in his ears now as he scented the nearby animals. As a young lion, he had had little to fear from hyenas, and he certainly never would have eaten them. But now, in this solitary state, his hunger was of a different quality and his attitude toward them had changed. Older, wiser, and more desperate, he might indeed eat hyena. Far back in his brain there lodged the glimmering perception that they might eat him one day. But the hyenas knew he was still dangerous, and they made no attempt to engage him. He heard them clomping along noisily out of sight, their heavy footsteps clearly audible in the oppressive heat of the midday sun.

Memories came back to him in falling asleep. As a young lion in a big pride, he had begun his own path to domination along the edges of the great lake, several days' journey from the plains. There, with three comrades, he had badly mauled an old buffalo. It had been an inexpert and clumsy attack which had forced the powerful animal to charge into the lake where he had turned to face the three hunting novices, his head lowered, his lethal horns swinging. The lions had splashed and leaped at him, lashing paws at his head, but they lacked the experience to get into deeper water and attack him from behind.

But the buffalo's wounds had been grievous, and these, not the lions, had finally brought him down. The lions had returned to the shore to rest and lick their wet fur. The young lion had napped while the other three watched. The buffalo's wounds had not stopped bleeding, and the water around him had gradually darkened as the thick arterial blood poured out from a dozen places.

He did not die in the sun. His last moments were reserved for a midnight moon when the three cats, lined up along the shore, had been at rest. The weakness of his back legs grew. The palpitations of his heart increased. His great horned head went down and within moments, he had drowned. But his death had not given the lions any special advantage. The young lion had swum out to the half-sunken bulk, the curve of the belly and the long horn protruding from the water. He had clawed at the body ineffectually. He was conditioned to regard the buffalo as one of the great sources of food, and he could not leave such a banquet to others. His comrades joined him and the four of them had ripped and growled at the carcass without effect for most of the day.

In the heat of the following day, the buffalo's stomach had distended with gases and he floated, a monstrous travesty of his living self, blown up to the size of a small elephant. But even this had not helped the lions because the wind blew offshore and the buffalo drifted away. The lions had followed, transfixed by the proximity of so much meat, while the buffalo, over two days, had become an almost perfect sphere, bobbing just beyond reach. Eventually, the wind had changed again, and the corpse came closer to shore.

The young lion made a quick, splashing charge and landed squarely with all four paws gripping the balloonlike carcass of the dead animal. It had been like trying to grip a rubber sphere. The weight of the lion could not bear the buffalo down. Both lion and buffalo twirled in a circle and the lion went completely underwater where he was jammed beneath the heavy body of the dead animal. The water was churned to foam and the lion struggled desperately to avoid being drowned himself. He shrugged his way clear, choking, gasping, and sneezing, and lurched through the shallows to the shore. Freed from the lion's weight, the buffalo floated away and was finally grounded on a peninsula of sand, where it was torn to pieces by vultures, hyenas, and storks, and its bones scattered along half a mile of shoreline.

Now, the sun shot up before the old lion, a perfect sphere of crimson flame turning rapidly to golden white, and the lion's amber eyes looked for long moments directly into the sun. A kaleidoscope of memories shuttled behind the expressionless eyes. He was a cub again, not even a quarter grown, and the rains had been torrential. His mother, that long-dead figure, was directly ahead of him at the edge of a river torrent. She tested the depth of the quick-running water carefully and then entered the river. The young lion and his fellow cubs understood immediately from her grunts and growls that they must stay close to her. Intuitively acting on her orders, they clustered along her left flank as she sank into the water. When the water deepened beyond their depth, they bumped against her side as she waded rib-deep into the quicker-flowing water. The young lion kept his head well up and showed his natural repugnance to getting wet, but he kept on swimming with all the trust of the young for the mother.

His mother began to swim herself and was carried downstream diagonally in the force of the current. She tried to keep her body at right angles to the current but with the cubs nudging strongly against her flank, her head swung downstream. She countered this with powerful strokes of her front paws but her hindquarters swung away. The last cub abruptly lost the protection of his mother's flank and was swept around her tail and downriver.

Feeling the greater thrust of the current and the alteration of his mother's angle of swimming, the young lion desperately increased the power of his own swimming. The second cub bumped hard against his hindquarters as he, too, responded to the danger, and the three animals finally reached the shallows. The streaming sides of the lioness heaved up out of the water and she paused while the cubs found their depth and splashed for land. She growled at them. Bedraggled, skinny-looking, and quite uncatlike, they went to cover and lost themselves in the shrubbery at the edge of the river. The lioness turned downstream and bounded from the young lion's view. He never saw her efforts to save the third cub. She swam desperately into the current and recovered him, gripping him by his hindquarters so that by the time he reached shore, he was dead, drowned in the rescue attempt.

While the young lion and his fellow cub waited for their mother, they heard stealthy noises. They stood rigid, looking through the tangle of stems that formed the shrubbery base. A large body was approaching, and the cubs shared the premonitional fears of their kind against all those who

62

wanted them dead—leopards, dogs, hyenas, and even some of the smaller cats. The cubs represented the world of the lions at its weakest point, the one time when the victim could strike back, and so their fear was real. When a large tawny nose thrust itself forward into the stems, both cubs spat explosively and leaped backward. Their fear and suspicion of another lion, if he were an alien, was just as great as it was for any killer. This lion, his scent now strong upon them, was an enemy, and they knew it clearly.

The stranger had heard the cubs running for shelter and suspected they were warthogs or young cats. He had decided to investigate, not because he was hungry but because he craved some sport. When he learned that his prospective playthings were young lions, he pushed forward confidently. They could not hope to stop him. His great face loomed over the young lion, jaws half-parted, and the cub had a scant second in which to act before he was caught. Instead of retreating as his brother had done, he attacked, and his sharp teeth penetrated the huge creature's tongue. In the chaos that followed, the uproar of the outraged lion's voice obscured forever the fact that a big paw had swept the youngster away like a fly being swatted from the flank of an antelope. He had been batted to safety while his brother had become the object of the lion's rage. Crippled, his hindquarters crushed, the brother had been thrown high into the air, then killed, and finally eaten.

The sun was now well beyond its midday point. Shining pink through a middle-afternoon haze, it bathed the trees around the old lion with lambent light that held stark acacias in gentle embrace and stole among the grasses and shrubs, coloring the rocks a soft rose. The sky was empty of vultures and eagles, and beyond the riverine trees the plains were bare of any kind of visible life. The peace of the moment meant nothing to the old lion, whose preoccupation remained fixed on the endless drought. He caught the fragile scent of leopard, but it did not engage his attention; he remained gripped in the fear of a dimly perceived premonition. He had no power to conceive of the finality of death, but he understood that the violence and death and injury he lived with every few days was approaching the center of his being. The pains that came to him during some of his careless kills cut deeper into his heart than before and seemed to lead to another state which, he knew intuitively, must be avoided. He dreaded its approach.

Even as a cub, he had experienced this dread and had sought to avoid the unknown state by any means available to him. After his escape from the lion attack, his mother had been absorbed into a large pride and he had joined six cubs, about his own age. With their endless fondness for playing, for baiting older cubs, for rolling and cuffing in the dust of long, hot days, they had replaced his two brothers in his need for company. He had become attached to a territory centered in a grove of trees which gave the shade the lion cubs preferred for their long daily dozes.

The pride had been so large that none of its creatures expected intrusions by other animals and were therefore unprepared for the arrival of a rogue elephant, a frenzied creature who had become insane when parasites reached his brain. The lion cub had awakened to see the elephant towering over him. He had fled with a growling hiss of fear. He was then too young to climb trees, and so he had run blindly ahead of the elephant, the earth trembling to the sound of the rogue's trumpeting and screaming. A gaping

64

aardvark hole had loomed ahead of the lion cub and he had made use of this lucky chance, bolting into it and falling to safety.

The hole had been deep and only recently abandoned. When the elephant reached it, his tusks had gouged into the earth but not far enough to reach the cowering cub. He had stamped the ground, gouged again, and trumpeted his madness. The earth had fallen in around the cub, the warm, choking soil muting the sounds of the insane elephant. Struggling for air in the damp darkness, he had not seen the elephant go completely berserk, roused to a frenzy by the scent of hated lions. The smell surrounded the elephant, oozing from the ground and coming off the trees. He had known there was a lion somewhere, and in his quest for the kill he had systematically knocked down every tree in the grove, torn up every bush and shrub, ripped up gobbets of grass until the glade was destroyed and nothing remained but his footprints and the lingering menace of his scream.

Buried deep underground, the lion cub had struggled to avoid suffocation. The original hole into which he had fallen had been thoroughly choked with earth, but because the aardvark had built in the remains of an old termite mound, as was his custom, the earth had been porous around the cub. Bat-eared foxes had occupied the burrow recently, and they had used another entrance. The lion cub had found this collapsed tunnel and had begun his fight to the surface, scrabbling the earth aside in a desperate effort to beat suffocation. The digging had been easy in the porous earth and he had burst to the surface in time to see the gray hulk of the elephant disappearing through the wreckage of the fallen trees.

The cub had looked around him. The pride had gone.

He heard no sound of mewling or growling, no grumbling of the old pride master. He had mewed, but there was no response. He had given a rasping cry of distress, but nothing answered. Then he had licked himself into confidence. His mother would soon be back. The other cubs would soon return. He had been lost before in the undergrowth. He moved into the shade of one of the fallen trees and lay down. Meanwhile, beyond his vision, his pride had been on the move, and such was the deficiency of the lion ethic that there was no memory among any of the lions, not even his mother, that he was missing. At that moment, he had been entirely forgotten, and only extraordinary good fortune later brought the pride back to the ruined grove, where he had been reunited with his mother. But in the long, hungry waiting, he had felt the premonition of that state beyond injury and pain, and the memory of it had goaded and guided him throughout the rest of his life. In every encounter with other lions, this dark memory had given him an extra charge of determination which led him safely through all the uncertainties of young lionhood.

Restrained by this memory, he had watched fellow youngsters charge with heedless lack of prudence at large and powerful victims. Still unskilled at the difficult tasks of coordinating the rushing momentum of heavy body, the swift swipe of paw, the clench of jaw and teeth, they went down under the horns of buffaloes, were thwacked by the double kicks of zebras, slipped on fresh new grass and dislocated shoulders. He saw their jaws break in vain efforts to crack the necks of creatures too large for their immature strength. He saw young lions led on by the seductive porcupine, their rage growing every second as they failed to penetrate that succulent body, saw them lose control and heard

66

the roaring fury when paws and muzzles were filled with stinging quills. He watched young lions die, their faces swollen obscenely while flies clustered thick and black on their distended muzzles; saw them killed in fights with older lions when they attempted prematurely to challenge the strength of the pride masters. And so, these dark memories a part of him, he tended to be the slowest to move, the last to charge, the most suspicious, and this had brought him through young lionhood into maturity.

The sun was high and hot, and flies were becoming bothersome. They concentrated on the old lion's left leg. Unless he kept his paw moving, they clustered there, sensing the corruption within and eager for a chance to deposit eggs. The lion watched the flies gather into a thick mass before jerking his paw away again and again. The flies were like the hyenas and the vultures of his old age: omnipresent and seemingly equipped with secret information about him. They knew, somehow, when he felt sick or tired, or when minor injuries made him limp or lethargic. The previous night an old hyena, the moon glowing in the white disc of her one blind eye, had come boldly close to him and had stood there, sniffing, while the old lion had rumbled a warning to her. Now, he had forgotten his confrontation three days before with the young lioness over the carcass of the wildebeest. In the fight, she had hooked a claw into his left paw. The momentary pain had quickly gone but the lion was not to know that something more than a claw had entered his foot. He growled softly at the paw and licked it.

For a year after he had been expelled from his mother's pride, he had been nomadic, circling the plains three times in rain and drought, and penetrating the woodlands five times in linear migrations that touched on all the possible territories he might one day settle. In these long and seemingly aimless wanderings, he had mated sporadically with lionesses he encountered near islands of rock on the plains, along riverine forests in the woodlands, and at water holes. These series of chance encounters had no feeling of continuity. Sometimes, he met lionesses with whom he had mated, but he felt no bond to them, no concern about their cubs or interest in them. But the encounters, although without apparent meaning, left him with a nagging sense of uncertainty, as if the sexual act had not, in fact, been completed.

At the beginning of his second year of nomadism, he returned from the plains and came across three lionesses, all in season, who were resting in the shade of trees along the dried watercourse at the northern head of the plains. In the next three days, he mated nearly two hundred times with the three lionesses, not pausing to eat or drink, although both food and water were available nearby. When he was finally done, and asleep, he did not truly waken for more than a day. The trees around him were silent. The grasslands were empty. The three lionesses had left the area. But the place in which he had mated had acquired significance. He knew it well enough. It was usually occupied by a pride, but now, he sensed the absence of the other animals. And it was in this place, eventually, that a lioness came to him, and then another, and they stayed with him because he radiated his readiness to possess a pride, and he did not move from this place, which became his territory.

The lion had placed himself in ideal territory with permanent water nearby, the plains accessible to the south and the woodlands to the north, so that in both wet and dry seasons he needed only to move north or south to find victims. He had managed to squeeze himself into country that lay between two prides and was accepted grudgingly by his powerful neighbors, each of them masters of four or five times the area over which he was able to hold influence.

Thus well emplaced, he had come to know exactly when the wildebeests began their march south, although he might not be sure which route they would elect to follow. He had learned the movements of the zebras, the positions of the main concentrations of gazelles, and where the elands might be quartered. He had watched for his enemies and listened to the rumblings of the other prides, trying to divine their intentions.

When he had stood alone in the near-total darkness of moonless nights, he had felt an urge to place himself in the center of his territory and to assert both his position and his power by roaring. He roared to creatures close and distant. He told them his whereabouts and his mood and warned them of his strength. He told nomads that he was in his hunting place and would not tolerate intruders. He told other pride members that the territory was well held. His roar was recognized by individual lions, since each lion's voice was slightly different from that of any other. But above all, his great bellows were an expression of his inner being, of his arrogance and confidence, directed at the heeding but invisible world around him.

But he did more than roar; he also listened intently to every sound. If he heard a woofing noise in the distance, he could be certain that one of his lionesses had been surprised; perhaps she had stepped on a snake or had been charged by a warthog. He might hear a cub bleating, a high-pitched and harsh nasal noise of extreme distress which signified that the cub was lost or starving. He listened to the grunting, choked roars of another lion signaling his position to his companions as he hunted. He heard the savagery of the cries made by lions when they killed, bringing buffaloes, or antelopes, or horses crashing to the ground. He listened for the return of his own great, full-throated roar, the loudest sound on the plains, and then he would respond to it. At close range, the air shook with the power of his cry. Preceding it came a deep, vibrating, rising moan, the prelude to the roar which, when it finally struck the air, was like a hammer shattering the plains atmosphere, and could be heard four miles away.

He was intensely sensitive to the bark of the socializing baboon. He understood the harsh, doglike cries of the bushbuck coming from thick bush, a call that usually meant the presence of a leopard. He listened to the deep chorus of crowing colobus monkeys which sounded like a congregation of birds in agony. He heard the sinister hissing intake of air when a nearby leopard filled his lungs before issuing a powerful, gruntlike cough of rage which communicated venom so completely that the lion had heard panic-stricken baboons dropping from their trees in response to it.

The lion's knowledge of sounds, although important, was perhaps subordinate to the eddying world of scents, of aromas, of powerful odors that swirled back and forth across his nostrils. The smell of his victims came to him from hundreds or thousands of feet away, depending on the direction of the night breezes. Sometimes, he could simultaneously pick up the scents of hare, jackal, gazelle, wildebeest, hyena, bat-eared fox, crowned crane, and guinea fowl. Every lion had his own smell too, and many of these he could recognize, the recognition stimulated by the habit all males had of marking territory or any line of march with their own scent. Each time he patroled the fringes of his territory, he detected his own smell or the smell of other lions overlaying his on trees or among clumps of grasses. He would stop and turn, squirting urine mixed with a liquid from anal glands at the base of his tail. This strong, musky scent lingered for days and told all other traveling lions and lionesses who had been there, and perhaps when.

As the years had passed, however, the lion had suffered from his aloneness as the patriarch of his territory. Invisible pressure to share the responsibility of this territory began to build. And because he, like all other lions, had a strong tendency to form a close friendship with at least one other lion, he was preconditioned to eventually receive another male into his pride. One night, after he had scented a newcomer approaching from the north, he had roared his customary warning and had received no reply. Nonetheless, the lion scent grew stronger and when there was no response to his second roar, he advanced toward it. The scent grew familiar in his nostrils and by the time the other lion came into view in the moonlight, there was no need for the newcomer to cry out or to attempt identification. He was recognized. He came into the territory as a former comrade, one of the juveniles who had worked with the lion on the shores of the great lake, attempting—unsuccessfully —to get at the swollen buffalo carcass floating in the shallows.

The scent of the newcomer was overpowering in the memory of the pride leader and so it did not matter to him that the newcomer had a convulsive twitch of his head, a sort of tic which shook him every few seconds, the product of some old disease or nervous disorder. The former comrade suffered too from an umbilical hernia which gave him excruciating pain every time he strained to jump or grapple with a powerful animal, or even when he tried to roar. This pain diminished his confidence during those times when the lions were roaring and prevented him from ever asserting any kind of authority during his solitary roaring. But it was precisely his disability that cemented the friendship, because he did not have the capacity, or the desire, to challenge the authority of the pride's original owner. Yet he was available to share the responsibilities of pride protection and dutifully marked out the limits of the territory with the same meticulous attention as his friend.

With his new companion, the lion saw his pride prosper. As it became larger, sometimes including as many as a dozen lionesses and an equal number of cubs and juveniles, it became more fluid in its composition and its rules of behavior were less strict. The pride members often spread out over the enlarged territory, now dominated by the two males, and perhaps might not meet for days on end.

But even when all the members of the pride were widely scattered, they still maintained the bonds of both the territory and the individual relationships between them. And it was their presence in the territory which restricted the number of lions who could hunt there, spacing out the hunting animals so that the amount of meat available at any time in the territory was much more predictable than if nomads constantly intruded.

Each member of the lion's pride shared the knowledge of where the best water holes were and the best ambush points along the river and in the thickets on its banks. The skill of the pride's hunting gave it cohesion. Every member enjoyed group security. The lionesses did not breed any more frequently than nomadic females, but they would rear five or six times as many of their cubs as those who roamed free on the plains.

The mood of the pride was precisely attuned to the determination of the pride leader and to the season in which it found itself. During the long, hot, dry days of drought, gazelles might be the only food available to the lions, and this made them all short-tempered. A gazelle carcass did little more than nudge the appetite of the average lion, and a gazelle kill disrupted nearly all the relationships within the pride. The lions went nearly berserk, and the tiny animal was briefly surrounded by a snarling group of creatures swatting at each other, biting and growling their rage at the small size of the meal. Then, the lion himself usually asserted his authority. In three out of four gazelle kills made by his lionesses, he drove them off and ate the carcass himself. In this dry season, he watched and listened intently for the leopard who was working part of the same territory along the river line, and took many of his kills away from the smaller cat. About once in every ten meals during the drought, he would make a kill himself.

With prolonged drought, the pride's mood became vicious. When food was abundant, the lionesses were good-natured, and the cubs played for hours in the shadows of the trees. But when the drought was fully mature and all the grazing animals had moved from the plains back into the woodlands, with only scatterings of gazelles walking through the territory, hunger created depression. The cubs became silent, no longer meowing or grunting their pleasure. They even stopped playing, and the mood of the pride became somber, anticipating the next outbreak of passionate anger at an inadequate kill. During this time, the pride would fragment, and perhaps half its juveniles and lionesses would disappear.

But when the rains fell, often in two great spasms, the first breaking the drought and arousing a gentle thrust of grass, the second torrential and bringing up the main growth, the lives of the lions were transformed. Then, all the grazing animals who had gone to the woodlands to survive the drought on the long, dried grasses there, began their great movement south toward the succulent, sweet, short grasses of the plains. Through the lion's territory poured wildebeests and zebras, small gazelles and elands, topis and kongonis. Buffaloes moved to the edges of the plains, and giraffes migrated like topless trees into an infinity of blue sky and green grass. The lions feasted and grew fat, and their numbers swelled to the limits of the territory.

Such prosperity dulled the group awareness of the pride and sometimes brought it to the brink of disaster. Once, the pride had been lying stupefied in the shade, having gorged itself on an eland cow killed two days previously. The pride leader was as deep asleep as any because he had eaten more than a just share of the meat, but he wakened to the sound of approaching buffaloes. He growled concern, then realizing their proximity, he roared a warning and began a self-preserving run for the nearest tree. His comrade male had joined him instantly, followed by seven lionesses and a straggle of youngsters who had wakened last and did not understand the danger. The buffaloes scented the lions and charged.

The sixteen cubs had not possessed enough experience to understand buffaloes. They ran after their mothers, but only a few could keep up with such quick flight. Some came to realize how serious was the danger and dived into clumps of sprawling grasses. Others ran into a gully and huddled together, whimpering at the sound of crashing buffalo hooves. The rest of the cubs managed to reach the trees which their mothers had already climbed and scrabbled desperately for footholds. But they were too young to climb anything. The leader of the pride saw two thirds of the cubs wiped out under the feet of the buffaloes.

Over the years, he became the repository of a great and seemingly disconnected knowledge of the lives of the other animals around him. On bright moonlight nights, he watched spring hares, the almost kangaroolike harvesters of roots and bulbs, and ground squirrels burrowing tirelessly in their perennial search for roots and fallen fruit. He heard the soft patter of the fast-moving caracal as it whisked through the darkness in pursuit of some desperate small victim. He saw the great martial eagles wheeling along the edges of the plains watching jackals during their breeding season, waiting for a chance to dive on youngsters drawn too far from their dens to return. He had seen the great eagles kill almost fully grown jackals and then, unable to lift them, provide the lion with a free meal. By occupying the edge of plains and woodlands, the lion was the witness and exploiter of all the travails of the north-south travelers. The golden jackals who roamed the great plains were pulled across the lion's territory at the beginning of the dry season when their victims left the plains. The black-backed jackals, creatures of the woodlands, were pulled in the opposite direction by the emigration of the grazing animals to the plains. Because the jackals were so precisely tuned to the seasons, the lion saw martial eagles searching the plains at the beginning of the rainy season for the young of the golden jackals, who bore their youngsters to coincide with the arrival of the grazing animals. During the drought, he saw the great eagles working deeply into the woodlands, where the black-backed jackals were giving birth to coincide with the return of their victims to their homeland, the woodlands country. At dusk, he listened to the howling of jackals, the high-pitched wailing occurring sometimes in duets and trios, an eerie and penetrating sound that traveled for miles.

The lion prospered on the predictable, the stable, and the certain. But not every year was the same. One year, a berserk bull elephant had treed nearly all his lionesses in a large acacia. The bull had thrust his weight against it, testing its strength, and then he had backed off and charged it. From out of the tree's branches had come a flying crop of desperate lionesses. The cats struggled to disengage themselves from the clinging branches, while the elephant, stamping and snorting and screaming, tried to tusk among the foliage to kill as many lionesses as possible. On that day, four of his females died, and it was two years before the pride recovered its old strength.

In some years the early rains did not fall at all, and the late rains were so long delayed that many of the migrant animals simply did not go out onto the plains. This caused periods of great confusion, with migrations and counter-migrations, the animals coursing back and forth across their trails with a baffling lack of sense or order. The lions feasted, the territory around them nowhere contested, then starved and felt the pressure of the powerful and hungry neighboring prides.

Such sharply fluctuating periods of plenty and scarcity

made all their tempers uncertain. During these times, the lion felt his mastery over his territory to be a less certain thing, and whenever his pride met its neighbors, roars and menacing behavior marked the contacts. His age began to influence him, adding to his doubts. There was an unaccountable stiffness in his joints when he awoke, a stiffness that became almost disabling if he attempted to run immediately after resting.

One night remained fixed in the old lion's memory. The moon was full, the barking of the zebras long since silenced with the horses gone to the north, and the droning cries of the wildebeests stilled by migration. His pride was spread along the banks of the quiescent river. From the south had come intimidating cries, and he had known that nomads were approaching his territory. They were all young, not more than four, and they roamed throughout the plains during the rainy season. They had established temporary territories wherever the hunting was best, and they thus acquired a sense of territory that they understood must include lionesses.

Hungry now, they approached the lion's territory with supreme confidence and aggressive intent. The lion and his male companion felt it, and both roared back their defiance at the challenge coming out of the dark air. The cubs and several of the lionesses fled. The pride leader advanced uncertainly, his comrade flanking him and slightly behind. Two lionesses moved out widely on either side of the males and confronted the five nomads. All the males roared. One of the lionesses attacked the nomads. It was a vicious, short-lived battle that sent her scampering away. The second lioness fled.

Then, for the first time, the pride leader had had a premonition of disaster conjured by a combination of his age and the richness of his experience. He had seen so much violence and death in the defense of territorial integrity. He had seen lions meet in river waters and, locking together, twist like mongooses. He had seen lions grapple upright in water holes, smothered in mud. He had seen lions flee through the trees in panic. He had seen lions lacerated in hide and foot, bleeding from ripped flanks and torn lips. He had seen lions with the tips of their tails torn off and ears ripped from their heads. He had seen their throats opened by lethal bites. He had seen how often claws caused wounds to become infected. He had seen muscles cut so that legs withered and the lions starved to death. He had seen lions so badly injured that their manes fell off. He had seen lions with their hindquarters paralyzed, and he had watched one lion, blinded in both eyes, blunder into trees, into thornbushes, into water holes and hyena burrows, meowing like a lost cub.

The five nomads had charged. The pride leader grappled and groaned, ripped and tore, saw his comrade go down under three of the nomads, felt bones breaking between his teeth, heard gasps of pain and passion. This, he understood, was no superficial conflict. This was a fight for survival. The killing bites, he knew, were to the throat, to the back of the neck which severed the vertebral column, or to the lower back which did not kill immediately but caused permanent paralysis and slow death by starvation. He twisted like a mongoose, got a lethal neck grip, crunched down his great jaws, and felt the neck snap. The body went limp beneath him. He drove off the second nomad, but when he scattered the three enemies attacking his friend, it was too late to save the hurt lion's life.

The pride leader licked his comrade's face, but he could do nothing about the ripped body. A large hole in the rump oozed blood. The skull was fractured and claw punctures almost covered the body. The lion lay down facing his friend and remained utterly still, watching. The nomads had not fled, he knew; they were waiting for another fight. His comrade tried to raise his head but the effort was too much. He groaned and gasped, his whole body convulsed by the spasm of air entering and leaving the lungs. As the pride leader watched, the spasms diminished, then ceased. The pupils of the expressionless eyes became very large. Then his comrade died.

The lion had remained by the body of his dead friend for hours. The breaking of this special bond left him uncertain, and the world around him suddenly seemed dangerous. The power he had expressed so confidently in his great, solitary bellows into the night air now seemed to dribble away into nothingness. He felt lost and imperiled. When he tried to roar, the sound came out as the harsh, nasal snarl of a lion cub in distress.

There was now no need to fight the four surviving nomads. His determination had gone, his territory had become meaningless. He turned and fled, fled over the bodies of the two dead lions, fled into a darkness deeper than that of any night. He had fled from pride and from territory, fled into country that remained alien to him for days. He was thus made nomadic by his age and was returned to the condition of his youth. But there was a difference which he could not at once comprehend; he would never again win ownership of a pride.

He had wandered aimlessly across years of rain and drought, wandered to the southern end of the plains and back again in great lazy circlings. He had no destination and moved only because his victims were moving or because the territory was occupied by other lions who would not accept him. He saw the plains with new eyes, often resting on top of a ridge, with the moon high amid serried ranks of clouds, and with dust hanging in the air and obscuring the sharp outlines of the animals before him. He would sit motionless and watch packs of dogs standing with their big rounded ears protruding while the passive wildebeests trudged in the moonlight. He looked out into long gray ridges of land leading to yet another plain, yet another ridge, yet another day, and so on into an infinity of nights.

He wandered amid winds that were sometimes so strong they ripped up clouds of startlingly white soda dust from distant dried lakes, and by their force darkened the color of the sky. Then, the contrast between the white soda dust and the blackness of the sky gave the plains a starkness and a barrenness that matched his own aching feeling of loneliness.

He walked in dawn winds howling up out of the west and blowing into the eye of a cold white sun rising above the plains. He saw hunting dogs lying in long masses, one dog's body protecting the next from the force of the wind. He was humiliated when he sought to join other prides unobtrusively. He was thrashed when he tried to hunt with a young nomad.

He was repulsed by lionesses he found alone. He was chivvied by hunting dogs, pestered by hyenas, and dogged by jackals. He walked a long and lonely path after his fall from power.

V

When the baboon awoke in predawn darkness, he could feel damper, heavier air which spoke to him of rain. The air lightened, the sun rose, the accursed leopard lay on his branch. Red light rolled across the plains and washed against the island. It dimmed under a band of low clouds hovering along the horizon. This line of clouds grew in the full circle of the baboon's vision, leaving only a narrow strip of blue sky. The island was covered by a dark cap of oppressive cloud. As the baboon sat in his tree, contemplating the risks of trying to dig out a hyrax burrow while the leopard slept, the rains began.

The rains—plural because they did not fall as a single

entity but in a series of torrents, drizzles, mistings—were spasmodic and unreliable; they could never be trusted. No rainy season was ever quite like another. The darkness of this day, alternating with sudden bursts of smoking light smashing out of the center of the sky as the sun strove to reestablish its dominance, indicated serious rain, reliable rain, heavy rain.

The rains were a magnet that drew the animals out of the woodlands, drained five thousand square miles of long-grass country of nearly all its grazers, a quarter of a million tons of flesh on the hoof. Willy-nilly, they must now run the gauntlet of the lion prides clustered along the fringes of the woodlands and the plains. Willy-nilly, they had to enter strange and dangerous territory, hundreds of thousands of the migrant animals making the journey for the first time with no idea where lions lay in ambush, when the hyenas might attack, creatures who might never in their lives have seen a cheetah.

The migration, which uncoiled itself with the deliberate gathering of strength that only such a massive movement of life could provide, brought revolution to all in its path.

The lions might kill and kill again, gorging themselves on antelope and horse, but because most of them were prisoners of their territories, it was a passing surfeit. Tribes of hyenas faced the same impasse, although they were not so rigidly bound by territory. The old hyena, anchored by her denful of cubs, would feel this because she could only travel twenty to thirty miles in pursuit of the migrants, and even such effort would be exhausting. Luckier were those lions who did not belong in prides because they had been forced out in fights for pride leadership or were too young to contest a place in a pride. As nomads, they could follow the game. With them moved old and unattached hyenas, jackals, and cheetahs.

The baboon heard the rustling of vultures' wings each morning, usually long before light occurred. He timed the birds' later departure to coincide with his first attempts to reach the ground. The leopard was most likely to be in his deepest sleep then, or at least disinclined to move. On the first day of the rains, though, none of the vultures left the island. They could not. They were as much prisoners as the baboon. They depended entirely on uprisings of warm air to carry them to sky-watching stations. So they waited, impatiently rustling their feathers and preening, scanning the sky and thrashing their wings in frustration when they saw other vultures who had roosted on high crags a hundred miles to the south passing overhead toward the woodland country.

The vultures who could fly widened their range to cover all the territory of the northern woodlands and the southern fringes of the plains. The migration was marked as boldly on the earth beneath them as if it were scratched by charcoal. The vultures saw the direction, the size, and speed of the migration. They watched the different coalitions of zebras converging on first growths of grass. They noted how many lions moved to join the zebras, how the resident prides were disposed along the fringes of the plains—this last information extremely important because these prides would kill much more than they could eat—how well concentrated the hyenas became in their pursuit. The rains were falling in a dozen different places within their eyesight, and they found odd updrafts between showers. They had such precise knowledge of the migration that a score of them were soaring over one place where a kill was expected.

The tsetses came as far south as they could, stopping as the woodlands thinned. They clustered in trees at the edges of clearings, unable to fly beyond them. They were inhibited by the rain, but their choice of blood was catholic. Warthogs now became their favorite blood meals, followed by giraffes and buffaloes, bush pigs and duikers, hippos and basking reptiles, rhinos and preoccupied birds. A hippo needed only to expose one flaring nostril to breathe in his pool when a tsetse would be clamped on the skin, the proboscis drilling for blood instantly. They drank less frequently from the active impalas and the secretive waterbucks. They had difficulty reaching the monkeys or baboons who sought shade and high resting places. The hyenas hated them and kept away from the woodlands in daylight. The cats, being nocturnal, were hard to find. And so, as the flood of life poured south into serried rain clouds, the flies turned back to the buffaloes who had retreated to day refuges, grazing and drinking at dawn and dusk. They concentrated on the predictable rhinos making their endless rounds through their territories, snipping twigs and browsing leaves.

The rains galvanized birds into movement harmonious with the new season. For some, it was the beginning of courtship and mating as the showers released legions of insects which had been quiescent during the drought. Mosquitoes hatched and matured in rain water caught in tree-hollow refuges. Spiders poured out eggs; beetles, bugs, wasps, bees, and flies rose from the damp earth in armies. Winged termites prepared to make quests for new homes.

Some birds flew thousands of miles to be at the grasslands for the rains. Others, particularly kites and buzzards, were animated by precisely opposite traditions. They deserted the rain area to breed in dry woodlands two thousand miles to the north. Hundreds of storks, wanting to remain in wet country, winged south from a rainy season ending in the savanna twelve hundred miles away. With them came a parade of others—hornbills pushing their long and heavy beaks in clumsy flights, kingfishers heading for familiar rivers and streams and ponds and expanding lakes, white-throated bee eaters who knew a product of the rains would be the flourishing of flowers everywhere. The colorful rollers also came south in swift plunging flights that caught the colors of their feathers in early morning suns and made them appear to be multicolored jewels speeding through the rain-dappled air. The journey to the feast of the rains went on night and day. The nights tinkled and churred and rasped to the sound of aerial voices rushing south. Owls feasted. Nightjars buzzed as they fled from lowland drought to highland rains. The movement of creatures were individual, by family, or by entire populations. Flights were regional, continental, hemispheric. Some came south to escape midcontinental heat; others headed north, where a few would die in late snow. They went east or west in search of more rain, or less rain, or no rain at all. They migrated toward higher altitudes, or lower altitudes. Because woodlands and plains were six thousand feet above the sea, and because the equator was close, every winged creature was within a day's flight of totally different worlds. Migrants to the eastern continental shore passed snow mountains buried in clouds and descended a long, sloping plain of thornbushes to a hot shoreline of palm trees. The western migrants flew over other mountains into dense rain forests that descended to a humid coastline. For those who went northeast, it was flight into an expansive desert. For those who went northwest, it was passage across a lush green land of immense swamps, waterfalls, and lakes where a large river was born.

Meanwhile, the island was smothered in rain, buried in downpour after downpour which drove the hyraxes out of sight and reduced the vultures to bedraggled blotches of black. The leopard disappeared. The baboon was ravenous. He sat like a vulture himself, wet and dispirited in the hiss and clatter of the rain. The plains had not yet sprouted green, and the mournful sound of water smashing down on puddled earth rang a dirge in his brain.

Beyond his sight, the greening of the plains unleashed the zebras, the wildebeests, the gazelles. The migration grew to the applause of thunder running from horizon to horizon. The crash of rains formed instant lakes, so deep and wide they forced zebras to gather at their banks, doubtful whether to ford or skirt them. The animals moved in herds and subherds, in family groups and small coalitions of family groups. The pace was gentle in places where heavy rains still had not fallen. It was tumultuous, almost frenzied, where the animals sensed that the distant rains had already brought up great growth. Like a broad, slow-moving river in flood, the migration was often temporarily dammed by obstacles, but it always gathered collective momentum and flowed over or around the barrier.

The sound of the migration was the whistling screams of zebras as they moved south in concourses twenty thousand strong. It was the honking and braying of lines of wildebeests strung out for twenty miles without a break in their ranks. It was the whisper of gazelles' feet touching muddied earth at dawn as small groups moved cautiously toward the plains. The migrating animals formed roadways and cut them deep into the earth. They created quagmires where they gathered at new pools to drink.

Along the routes of the migration, many kinds of disorder prevailed. Lions holding territory in the paths of the migrant could not eat all they killed. The old lion did not even need to guard his latest kill but could sleep near it certain it would still be there when he wakened. The grazing animals obeyed the orders of tradition and took their customary routes south, regardless of the conditions. A narrow roadway used by gazelles ran through thick brush between two towering slabs of rock, and consternation spread when the tiny animals found it blocked by bloated, sleeping lions. The gazelle marchers could contemplate no other route, and they backed up for a mile or more in milling indecision. The leaders were made irresolute by the sight of their supine enemies and by the pressure from the thousands who did not understand the delay when they saw no reason for it. Eventually, the tension became too great, and willy-nilly, the gazelles charged. It was a shaky, terrified advance of the timorous pushed by the cumulative power of the silent majority behind. Tails flailing, tiny bodies leaping, the gazelles pressed forward into the narrow pass. The lions were trodden on and woke at once, grumbling guttural disapproval. Several got up and batted down the jumpers. But in such a concerted rush of animals the lions were overwhelmed and went down under their weight. With choked roars, lions struggled to escape, to run or crawl from the path of the meek suddenly become mighty. The mass of gazelles went through the pass, smashing the undergrowth flat. Dead gazelles lay where they had fallen; the ground was muddied with the imprint of hundreds of thousands of footmarks, and the lions were hidden and watchful.

A group of wildebeests reached the shores of the swollen river at the edge of the plains and paused there. The river held no crocodiles, but lions were visible, lying down on the farther bank, and the leaders of the wildebeests saw them. They vacillated, their moaning voices swelling as they were joined by a second, and then a third, large herd. The lions yawned.

Shortly before midday, a dark-skinned bull, almost black, led the first group of animals down into the water. They snorted and shook their heads against the current. The hinder wildebeests pressed forward at a gallop and the animals ahead felt themselves being overwhelmed by the superior mass and speed of those behind.

The lions, awake now, well fed and rested, watched bemused while the wildebeest leaders became a conglomeration of black bodies folding into themselves as they went down into the water. There was no escape from the pressure the wildebeests had created. While the leaders thrashed in the water, frantic animals climbed on their backs as they sought to avoid the fate of those beneath them. The lions stood and looked down at the honking carnage in the water. The river, partially dammed by the fallen, had increased its flow behind and ahead of those who had gone down. Many who sought to clamber over the island of the fallen were swept away by the increased force of the current. Then a rip current developed, sweeping many more off their feet.

It looked like triumph for the river, but there was no way

to stop the crossing. The wildebeests poured over the rise of the bank and swept down into the water. The force of the collision with the river became unimportant because the wildebeests were now crossing unimpeded. The fallen had created a bridge over which the river poured. The animals lurched and swayed from one treacherous foothold to another along this wildebeest bridge. Urgent animals, still led by the black bull, streamed past the lions, eyes staring and black manes gushing water. Many did not even see the lions. When the last of them had passed, the lions were still looking at the tumbled pile of bodies disintegrating before their eyes. The half-drowned rolled downstream and lurched to dubious refuges along the banks. The drowned turned belly-up or lay where they had died.

The rains were now general and had drenched the plains from woodlands to the edge of the hills, from western lake to the thorn forests of the south. The speed of the grass growth unified. The inpouring animals left their roadways and quickly fanned out to graze the unlimited grass ahead. It was a rush to reach the sweetest grass first, to get to pastures uncontaminated by the dung of animals ahead.

The baboon remained on the island. The island floated in a sea of new grass, pallid green in a smoking dawn; thick, rich green by noon. Freshly pupated beetles appeared among the island's sprouting grasses, and the baboon ate them. He dug in the soft ground for insect eggs and nibbled sweet new shoots of grass. He wedged himself between two slabs of rock and blocked a tunnel, trapping a group of hyraxes who screamed and whistled pure panic. Their burrow was not quite deep enough to prevent the baboon's getting his arm almost to its terminus. He stretched down and could feel a tumult of animals. He moved his hand quickly to avoid being bitten, seized the head of one hyrax and pulled the screaming animal out of the darkness.

That evening, after he had eaten and settled into his refuge tree, he heard the migrant animals approaching. A mutter of sounds, of honks, hoots, whistles, and growls, was followed by a more general sound, like the soft roaring of water rushing over rocks. It was the noise of thousands of moving feet and thousands of chomping jaws.

The zebras, like all horses, had teeth on both upper and lower jaws. The other grazers possessed top teeth only. The zebras, therefore, led this army of the migratory thrust toward the long, ungrazed grass which they could bite efficiently. Behind the zebras would come the wildebeests, although this was not always predictable. They would take the grass down closer to the earth, leaving an inch or so uncropped. Finally, the gazelles would arrive. Their tiny mouths gave them the capacity to nip almost to the root-stock of the grass. In this way, parts of the plains were trimmed to a uniform height.

The zebras moved forward slowly, making room for the wildebeests. Groups of five and ten thousand wildebeests gathered in coalitions of one hundred thousand or more where heavy rains had brought up another flush of growth. The uproar of their voices continued day and night, and the background roaring of lions was muted to an inconsequential murmur. So many wildebeests surrounded water holes near the island that drinking became competitive. The panic-prone wildebeests charged forward, trampling their fellows into the mud. The baboon sat hunched in his tree and revealed no emotion as he watched the extraordinary spectacle around him. With the wilde-

90

beests came attendants: nomad lions pacing the migrants, hyenas who had no territorial restrictions, the wandering hunting dogs, jackals, vultures, and storks.

The moving animals passed among crowned lapwings
attempting to breed. Lapwing nests were studded across
the plains for more than one hundred miles. The wildebeests were careless where they walked. They were in-
clined to sudden bolts in groups of several thousand, panicked by hunting cheetahs, leopards, lions, or hyenas. In this
uncertain world, the lapwings defended their nests by displaying the white insides of their outspread wings as warning flags. In the low light of late afternoon or early morning
the sun shone on the white wings, often giving just enough
warning to turn aside the moving wildebeests. But at night
when the hunter and the hunted raced across the plains
world of the lapwings, there was no defense. The birds,
night senses poor, remained at their nests while the ground
trembled under the heavy feet of thundering dark shapes.
Lapwings and their nests were crushed in a world over
which they had no control.

The growth of new grass set many other kinds of birds to
nesting. Larks began nests in the thick new growth on
every low-lying piece of ground where ponds were forming
in the rains. The wildebeests trampled them. Beautiful
crowned cranes sat on nests openly displayed beside the
new watering places. The wildebeests crushed eggs and
nests in their blind progress to the next pasture. One morning, the baboon awoke to silence. The wildebeests had
gone.

But such was the dynamic thrust of life sponsored by the
rains that not even the heavy use of the grass around the
island could long delay its own powerful push. Within two

91

days fresh shoots surrounded the island with a dainty new patina of pallid green and, responsively, the gazelles appeared. They, too, arrived in large numbers, a coalition of several large herds united in the common goal of cropping the broadest possible front. But with the gazelles surrounding the island, the baboon was even more thoroughly trapped because of the large numbers of attendant hunters. Two cheetahs were visible to the north, slumped down as though they had been there all the time. The baboon could see several hyenas slouching through the grazing animals, neither hunter nor hunted disturbed by such proximity. A solitary lion stood motionless south of the island. Odd jackals trotted among the gazelles, waiting for that moment when a hunting attack would be made and they would be able to scavenge scraps from the feast. The scene appeared peaceful, but the baboon understood this was a lie. He was trapped. He had no organization of comrades around him to fend off the hunters. He was the ultimately vulnerable creature, unable to venture one step from the island without arousing the lethal interest of almost all the visible hunters who understood his vulnerability.

He could see food everywhere. His sharp eyes caught the movement of hundreds of creeping, crawling, and flying insects, ranging from spiders darting through the new grass to emergent millipedes encouraged to hunt in the heavy growth. But the most tantalizing of all the foods now starting to parade before him were the termites.

Light rains did not arouse the termites. They understood the unreliability of the rains, and before they would move they had to be sure of thoroughly drenched soil and a certain flush of growth anywhere they might fly. Their knowledge of the rains was so subtle that some years heavy rains did not arouse them either. They would be proved correct when, after the initial heavy downpours, the rains stopped entirely and the landscape dried again. But this year they were ready to move. One morning while the baboon agonized in his tree, they moved, the migration precisely synchronized from one end of the plains to the other. The day dawned clear. The air was calm. The workers in the termite cities climbed to the tops of their mounds.

This was the moment of supreme danger. They proposed to pierce their fortresses so that their winged progeny might be launched into space. Each worker drilled the smallest possible hole, close to the others, so that the top of the metropolis was punctured by scores of tiny holes. Because of the high risk, the warriors were right behind the workers. They actually emerged first in places and stood on the tops of the mounds with their large jaws open and their antennae blindly probing the air for signals of danger.

Workers poured out into the open. Their charges, who had been led from resting and caring chambers deep in the mound, came close after them. They were the kings and queens of the new civilizations to be created. For long moments confusion possessed many of the mounds. Some of the winged creatures were reluctant to leave and had to be pulled or pushed out of the holes by workers who scurried about in near hysteria. Other kings and queens were so eager to get out into the open plains that they had to be restrained by workers or warriors. For reasons of control as mysterious as all the functions of termite cities, some were pulled back inside the mounds. Kings and queens scheduled to leave on later flights could not change the order of departure.

For the baboon, the appearance of the termites became

exquisite torture. Like almost every other plains animal, he loved termite flesh. Yet it was a tantalizing and frustrating food because he could never catch enough of them for a satisfying meal. Countless times, especially when he had been younger, he had torn termite mounds to pieces, only to find his fingers probing into a chaos of collapsed galleries and passageways, the termites buried in the debris. He picked in the ruins and ate a score or so of the fat-bodied creatures. The warriors defended their cities and sacrificed their lives by painfully biting his hands and body, his lips and nose. He could never know that the bulk of the termites had fled underground into refuge galleries deep enough to be always beyond his reach.

But he had learned enough to know that when the termites emerged in a thick enough stream, he could hunt them in midair. Years before, when he had been a juvenile at the fringes of the baboon hierarchy, his troop had come in brief contact with a pack of chimpanzees. The chimpanzees had demonstrated to him how to hunt termites, but because he was still millions of evolutionary years away from conceiving the idea of a tool, he merely looked dumbly at the work of the chimpanzees, understanding only that they were somehow able to get termites out of their mounds. Chimpanzees inserted slender lengths of wood into the mounds and held them there until the wood became clustered with furious warriors biting into the stick in efforts to expel the intruder from their city. When the chimpanzees withdrew the sticks, the warriors remained inflexibly fastened and stayed so until they were licked off and eaten by the apes.

The baboon understood nothing of this, but now, seeing a particularly heavy outflight of termites from a nearby mound, he scampered over to it. The flush of insects was so thick that he could grab them from out of the air, even bite them in midflight, the insects packed so thickly that he got almost satisfying mouthsful. All around him, meanwhile, winged swarms were appearing from the tops of the cities. Many of the flights were poor and straggling, and fell to earth. Others flew toward the island and blundered about before falling. The baboon noticed that the lion's attention had been diverted by a bat-eared fox scampering among the gazelles. He was almost certain that the leopard had not returned, although the big cat could conceal himself with uncanny skill. He took a chance and continued to eat the unexpected meal of termites.

The termites flew into a hostile world. Their swarms attracted birds like bees to honey. Scarcely any bird on the plains or in the woodlands ignored the appearance of the termites. They sought the insects in insane, gyrating flights. When the termites chose to land, another host of hunters awaited them: the ubiquitous ants, eager lizards, hunting spiders, frogs, snakes, praying mantises, beetles, and hunting wasps. A flying feast was spreading across the plains, and the banquet, all the hunters knew, would be short.

Even as the slaughter progressed, the queens were landing. Few of them had traveled far, but the termite cities were already so well packed together on the plains that the flights tended to intermingle. One column of insects headed north through another group heading south. The baboon's keen eyes watched the landing of one queen near the island. She landed in the middle of a footprint of the hated leopard. She shrugged her body and her four wings fell off. She had found her place. But now she needed a king. Slowly, she raised the tip of her abdomen and forced out a penetrating perfume which rose almost vertically in the still air. It was quickly scented by a winged king flying overhead. He dropped down to join her and just as quickly shed his wings. Working together, the two insects dug a hole and disappeared into the soft earth.

This expansive tableau was repeated millions of times throughout the plains. Each of the king and queen groups was destined to work without regard for the fact that most of them had chosen, perforce, unsuitable city sites. Few of them had much chance of survival in a world where termites were the most numerous creatures on the plains. If it was an exercise in futility, it was also an experiment in grandiose ambition. Success depended on the queen-king combination's digging near an old or failing city or chancing miraculously on an ideal place for a new metropolis. Millions of lives must be sacrificed to perpetuate a few thousand. That was the order of the termite migration, and it possessed its own justice. The starting of new colonies was less an attempt to add more termites to a plains world already teeming with them than it was a testing of the prime qualities for future colonies. The new must be matched against the old. The queen and her consort must pit their genetic determination against the others. This selection began the moment the queen had landed. It was decided by where she chose to dig and by how deep she burrowed.

The queen who had disappeared into the leopard's paw print was fortunate. The ground there was slightly raised, which reduced the chances of flooding. The soil was neither too light nor too sandy, so there would be less chance of the new city's burning out in drought or being swamped in a glutinous mix in the rains.

The young queen faced enormous odds in getting her colony started. Aided only by her consort, she could not burrow far down. Her first birth chamber would be within easy digging reach of the many who esteemed termites. It might take her five years to establish the nucleus of a colony. She would be long dead before any great metropolis arose over the tiny shelter she and her consort were now digging.

As the baboon worked away at his termite hunting, other creatures were emerging from the soil of the plains and woodlands. Tiny towers sprouted fresh earth nodules. This was the harvester ants at work. Like the termites, they had been stimulated by the rains to expand their networks of underground living quarters where their harvests were stored. They were working day and night to gather a fresh harvest from the surface of the blooming earth. The ground beneath the baboon was riddled with tunnels connecting colony after colony. They stretched around the shores of the island, satellite cities connected to one immeasurable metropolis, the network bespeaking purpose that made harvester ants models of organized industry. Unlike their cousins in the termite cities, the workers and soldiers of the harvesters could see, as they must, since much of the correct crop material had to be selected visually. But if eyes were not needed, the harvesters could be blind, and some, indeed, were blind, working only at night to gather crops. These blind harvesters were cautious. After digging tunnels to the surface and harvesting above ground, they sealed the tunnel holes when they returned underground. No bird or mammal might know they had surfaced and returned to their subterranean homes.

As the density of grazing creatures around the island

increased again with the arrival of fresh contingents of wildebeests, the industry of the termites was often trodden to rubble, and the termite workers were kept busy rebuilding.

Centuries of excavation by the termites had given mongoose packs easy refuges. Almost anywhere in open country they could quickly create burrows in termitaria. They never occupied any of these dens for very long. Sometimes they dug into working termite cities, causing chaos within the mound as the termites relocated their operations deeper underground. But because of the high mortality rate of older termite cities, it was usually easy enough to find dead cities, even though the mongooses were in competition with bat-eared foxes, hyenas, and jackals. The termites tended to strip away all vegetation around their mounds and this suited the mongooses, who found vegetation near their dens uncomfortable.

The arrival of the new wildebeest hordes brought migrating mongoose packs with them and galvanized the resident mongooses who had lived around the island during the long drought. The baboon watched them too. When he had been in his troop, mongooses had often led the baboons to feasts of insects. A favorite food of mongooses was the beetle which appeared in millions to hunt the dung of the migrating grazing creatures. The mongooses followed the tracks of the grazers where dung lay thickest and picked off the beetles as they worked along the tracks. The beetles partially avoided the mongooses by working at night. But like the termites and harvester ants, they were under pressure to get as much of the dung collected and buried as quickly as possible.

The mongoose packs left their burrows shortly after dawn and soon split into two groups, each composed of about twelve adults and youngsters. In the den remained the very young mongooses, a dozen or so, too small to run with the adult packs. A single mother guarded them.

One of the split packs headed along the roadways of the grazers, while the other branched away from the tracks to forage for random dung dropped by grazing animals rather than migrants. They worked methodically, well spread out, so that each animal was sure to get his fair chance to check out every piece of dung. Acute powers of scenting and hearing guided the mongooses, and they paused before each pile of dung to listen intently before quickly breaking it open in search of the beetle they knew had already burrowed inside to eat or lay its eggs.

The beetles were equally thorough, but they lacked the speed and mobility of the mongooses. On the stealthy approach of a mongoose, a beetle about to enter a dung pile would drop to the ground and make furious efforts to dig under the pile, or insert his body in a crack in the ground, or burrow out of sight before the mongoose could seize him. The mongooses' power of digging was much greater, however, and most of the beetles thus found were eaten.

The baboon, unable to stand the temptation any longer, unable to see the leopard, dropped from his tree and moved cautiously out into the plains where insect and mammal struggled. Many of the dung beetles were tiny, scarcely more than a morsel in the mongooses' small mouths. But uncountable thousands of them swarmed, frantic to escape their worst enemies, yet preternaturally bold to be abroad in daylight. The baboon worked with the mongooses. They ignored him. He found dung beetles pushing carefully built-up balls of dung toward burial places and picked them up and ate them. The mongooses competed with ants in their hunt and if the opportunity offered, they followed the ants back to their nests, dug into them, and ate out all the cocoons.

In two days of hunting the baboon ate more than twelve thousand beetles. Despite this, each day brought growing numbers of dung beetles. The thickest concentrations of dung attracted the largest beetles. In one place ten thousand of them worked to dispose of a mixture of zebra and wildebeest dung before it dried out, blew away, or was washed back into the soil. The amount of dung was not adequate to supply this number of beetles, and so baboon and mongooses came on a kind of battle royal among a dozen species of dung beetles fighting each other for a share of the booty.

The mongooses' excitement was expressed in soft calls— *merr-merr-merr*. Lithe and efficient killers, they jabbed at the stock of food now so easily available with forepaws, mouths, and bodies thrusting at the dung. Now and then one of them stood up, perfunctorily checking for danger, but the banquet was so great and the greed of the mongooses so strong that little attention was given to possible danger. Their hunting did not stop the fighting among the beetles. Beetles gripped in combat were oblivious to the presence of either mongoose or baboon, even as they were caught and lifted into mouths. Beetles died still kicking aside competitors for this mountain of food.

One midmorning, the sun blazed hot, and the baboon, his caution blunted, strayed hundreds of yards from the island to hunt for dung beetles among the surfeited mongooses. A watcher oversaw the feast and knew its meaning. The baboon, standing so much higher than the mongooses, could watch for lion or leopard and he was not overly afraid of cheetah, but the mongooses were not so well equipped to detect danger. They did not see a martial eagle wheeling. He was almost invisible because of his height, but his telescopic eyes recorded the scene below. He knew all about the preoccupation of animals feeding and knew what to do about it. A monkey possessed by a favorite nut, or a wading bird among his favorite aquatic plants—these were victims easily caught. In a spiraling drop he slanted away from the mongooses so that he would fall behind the island.

The baboon saw him just before the eagle disappeared. He hesitated, wary of the many connotations of his exposed position so far from cover. He had seen the big eagles take young chimpanzees and monkeys. They were dangerous. He understood, too, that the fall of any large bird would be marked by other hunters. They might be turning toward the island already to investigate whether the eagle had been the observer of a kill. As the invisible eagle was dropping behind the island on its eastern side, the baboon galloped toward it from the west, each hidden from the other.

For the eagle, flattening out at faster than one hundred miles an hour, the difficulties of any attack on mongooses were predictable. He understood their vigilance, but he had such good cover between him and his prey that the attack seemed an excellent opportunity. The only doubtful issue was whether he should pause at the island and risk being seen while he roosted, or whether he should plunge straight around the island and assume that he had correctly placed the position of the feeding animals. He elected to make the direct attack. He aimed himself to pass along the northern shore where trees would give him cover until he burst into view.

But as he shot around the northern side of the island he found that he had misjudged the position of the mongooses. They saw him immediately, saw his wide flat wings turning him toward them. It was too late to scatter because his speed far exceeded their capacity to find cover. Instead, they rushed together and formed a compact ball of bodies. Half-raised, they all faced the onrushing bird. His attack was not to be a strike at one relatively helpless animal, but a charge into a bunched mass of enemies who would not hesitate to leap at him. Such collective defense sometimes pulled down flying attackers. The weight of mongooses fastened to all parts of a hunter's body forced him down, even though he had made his kill. Down he came with a dead victim in his claws, to be eaten by those whom he had expected to eat.

Despite his size and power, this martial eagle was made cautious by age. He had already experienced the collective defense of monkeys and mongooses. Although he maintained the angle of his attack until the last moment, he had abandoned his intention of striking directly into the corporate body of his victims. They did not panic. He did not swoop. He sailed overhead at high speed, legs and claws retracted. Nonetheless, one of the mongooses, bolder than any reason, vaulted at the eagle and almost got a grip on the big bird. He tumbled to earth with a mouthful of feathers. The mongooses scattered so quickly that by the time the eagle had begun his turn back toward the island they had disappeared into the earth. The last leap had been critical. It had dismissed all notions of a mongoose meal. The eagle did not look back at the sparkles of light in a score of sharp eyes watching him from the earth. He landed clumsily in the highest tree on the island and was disconcerted by a black form hurtling upward at him, as if in attack. The baboon did not immediately see the eagle. He paused and looked up. While their eyes met, the leopard raised his head from the branch of his tree near the baboon.

VI

The old lion, as was his custom, had followed the grazing animals onto the plains. He had been held at the edge of the woodlands by the concentration of animals building up briefly here and there before their streaming advance to the plains. Thus, some of the migrants were already seven to ten days deep into the plains before he began his own migration. Behind him, scores of other nomads, well spaced out, and moving singly or in groups of up to six, followed the grazers at their own pace.

In his pursuit of prey, the old lion would cover up to fifteen hundred square miles of territory in the next one hundred days or more as he criss-crossed the plains in search of the best hunting. If he were lucky, he might fall into a temporary association, almost a friendship, with another nomad, most probably an older lion like himself, and the two might even be able to establish a more or less fixed territory, where pressure from the other hunters was light and the presence of victims was heavy. But in all the experience of his years of wandering, such an association would mean an exceptional year, and he did not anticipate it.

The old hyena, meanwhile, was being pulled south by an even more precise magnet. Unlike the lion, who casually followed all the grazing animals, she was tied to the march of the wildebeests, and because they moved in such dense masses, taking longer to spread out than the other grazers, her position on the plains tended to be fixed by their placement. This meant an immense elongation of her own territory, if now it was indeed territory, with its base still the den where she had her cubs. In this elongation, her associations with nearly all the other hyenas she had hunted with in the woodlands now changed drastically. Some of the older animals dropped away from the tribe; strangers joined it. An almost constant series of meetings took place as nomadic hyenas from the plains attached themselves to the tribal creatures and those who had spent nomadic periods in the woodlands. This was a time of fluid territory, of great uncertainty, and of a constant testing of power and dominance. Only one thing was predictable; each night of her hunting, she was drawn gradually further and further south into the plains.

On the island, the baboon looked up at the martial eagle and froze. The sight of the unexpected bird almost destroyed the strict balance between caution and panic he had managed to maintain in this new and hostile world. But the old eagle, in no mood to meet baboon after his near escape on the plains, took off from the tree and in moments was out of sight.

To be trapped on an island surrounded by a sea of food was dilemma enough for the baboon, but a larger pang was his separation from the troop. Like all gregarious animals he was married to the group, bonded to it so completely that no part of his personality dictated that he should be a loner. His physical makeup, his adaptation to territory, season, and aggression were shaped in the group. The terror of the nights, the need for the comfort of comrades around him, combined now to make him depressed and hysterical by turn. Daylight made him wary because he was not safe in trees or on the ground. He remained a prisoner of strict territory, of familiar places visited again and again. The visits were timed to coincide with the ripening of fruits, the maturation of grass seeds, the swelling of plant roots, the laying of eggs, the ripening of seeds in trees, the passage of migrants, the moments and places where helpless creatures were born. These were some of the imperatives governing the discipline of territory, all modified by other baboon troops who might share or contest parts of the territory. Uneasy truces prevailed at water holes, but during food shortages no territory could be called secure, and conflict became a necessary part of survival.

Territory was the predictable, the anticipated, the repetitive, the secure, the safe. Denied these ingredients, the baboon must remain in limbo as he watched dawns and sunsets, migrations and rains, hunters and hunted moving around him in an alien world without baboons. Worst of all, he was held on the island by a creature who shared none of his gregarious feelings, who moved unpredictably, and whose territory, if he possessed one at all, was ill defined. The leopard became a dark presence that haunted each second of awareness.

The baboon was a captive of the moment, not well able to resolve the unexpected. He was now, in fact, less able to survive than the outwardly defenseless grazers, though they, too, were prisoners of their own systems of life.

The gazelles appeared to move in large concentrations, but were separated into small groups dominated by a male. One buck gazelle paused at the edge of the river, now flowing strongly, and snorted his fear of both the crossing and the scampering that he knew lay ahead in the undergrowth on the far side. Of all the migrant animals, he and his kind were the most vulnerable, weighing sixty pounds or less apiece, their bodies nowhere near as husky as those of the larger gazelles permanently resident on the plains, their horns sharp but rarely used in defense, their only strategy the capacity for rabbitlike runs from their many enemies. The buck's great strength lay in his ability to procreate throughout the year, this ability enhanced by the short gestation period of his does. The gazelle numbers were built up far beyond the capacity of any meat hunters to noticeably depreciate them. Snorting again, the buck put one hesitant foot into the water. The most critical part of his migration was about to begin.

The baboon looked at the gazelles around his island. In his good times, he could seize a young gazelle hiding in the grass and disembowel it with one deft motion by sinking his massive teeth into its belly and then using his arms to wrench the body away from his toothgrip. This tore the animal in half. His teeth, extraordinary for a herbivorous animal, were dangerous enough to slow the leopard. But by comparison, his bowed, short legs forever prevented him from having the great range and freedom of territory of the frail grazers around him.

Scuds of rain washed the plains into indistinct distances. The grass, cropped by thousands of nervous little mouths, was being taken down to earth level again. Soon it would be gone, and the gazelles would move on. They seemed to wander freely back and forth, but in fact, they could only move where there was grass, where there were not too many other gazelles, where the zebras and wildebeests had grazed the higher grasses ahead of them, where there were no heavy concentrations of lions, not too many groups of hunting cheetahs or dogs. They were locked in a territorial prison, wider than the limited shores of the island, true enough, but just as rigid as the baboon's own trap.

The baboon's eyes lost focus on the gazelles and his gaze became fixed on that inviting horizon again. His sense of time and place on the island disappeared as the great longing possessed him once more.

The buck gazelle, searching for a sign of the leopard, looked up into the rocks and trees and shrubs. He saw the baboon. He snorted and stamped his forefeet, appearing to challenge the baboon to leave the shelter of the island and attack. This was now his territory, at least part of it around the island, and as a leader, he sought to establish as much authority as possible over it for the time he would be placed here by the push of the season.

Although he was frequently confused in his reactions to challenge, although his territory was often ill defined and almost constantly changing, the buck gazelle had that special quality of survival which marked the most successful territorial animals. During the long trek down onto the plains, the gazelles had been attacked a score of times. Each challenge had to be met with a precise response to each of the different hunters. When a pair of jackals came among the grazing herd during its second day beyond the savanna woodlands, the buck and his nearby colleagues simply ignored them. The jackals were not powerful enough to attack adult gazelles. Although many of the does were pregnant, fawning would not begin for weeks, and the jackals were no threat until then. Some slight risk existed if one of the gazelles had been injured in a night attack or had become sick and was in hiding trying to get well. But the chances of the jackals' finding these victims were small. They were usually discovered by the omnipresent hyenas, by watchful leopards, or by young lions.

The gazelles had come steadily south in the grip of sweeping rains, though sudden, quick spasms of drought and heat sometimes held them uncertain. Throughout the migration, the buck gazelle fought every step of the way to establish territory. He fought for territory at a water hole, even though the gazelles were not drinking—they scarcely ever needed water. He locked horns with other buck gazelles to ensure his modest territorial demand. They pushed and shoved until one buck broke away and trotted off. Neither animal was hurt in these encounters. He fought for territory along the banks of the river flanking the woodlands and at the edge of the plains themselves, and now, deep into the plains, he was still ready to fight his colleagues for featureless pieces of earth.

This was easy enough, a simple series of decisions to make, but the decisions became complicated whenever a hunter appeared. An old female hyena had come clumping into view one midday, walking among gazelles grazing and gazelles resting. Her udders hung low, and with her head-down stance and her heavy footsteps, she was making no effort to conceal herself. The buck watched her warily but divined that she was not dangerous. He resumed grazing. The hyena passed so close that she could have rushed forward and seized him easily enough, but she disappeared, passing clear through the center of the unconcerned herd.

In the late afternoon of the same day, though, a pair of hyenas had appeared from the north, their rounded ears on the horizon announcing their approach. Instantly, the buck was alert. All the other bucks sensed his concern. Every head came up. The buck advanced a dozen hesitant steps, snorted, then held his ground. This was known danger, not just because there were two hyenas—hyenas frequently hunted alone—but because the quality of their interest was totally different. Now, the shape-up would begin, the hunters estimating the wariness of the gazelles, while the gazelles counterestimated the danger of attack. Instantaneous flight from all possible sources of danger was wasteful and time-consuming, and compared with the number of animals killed in any series of attacks, not worth the effort. Besides, the hyenas—and most of the other great meat eaters—frequently did not attack.

Normally, the buck would have allowed the hyenas to come within five hundred feet before turning himself to flee. But something about these two hyenas—youngsters hunting together—communicated their hunger to him, and the buck prepared to react correctly. He waited to find out whether they intended to approach steadily or whether they had decided to make a rush. His response to a steady approach would be one burst of flight before he turned and faced them again to reappraise their intentions. They rushed.

Instead of fleeing, the buck began dancing. With legs stiff and neck bowed, his dancing was an odd up-and-down movement that held him in one place. The dancing, or stotting, or pronking—there was no adequate word for it—spread instantly to a hundred other gazelles. An undulating mass of brown backs and striped sides leaping insanely up and down became the unbelievable backdrop to the attack of the two killers. The inexperienced hyenas did not understand that they had already lost the hunt. They ran at full speed, confident of a kill.

Magically, the dancing stopped in relays, and the tiny antelopes darted away one by one. Their flight, once begun, was many times faster than the top speed of the hyenas, who were too young to know that gazelles rarely ran very far and were not capable of long-distance flight. As the attack was pressed forward, it became apparent that the hunters had not made one crucial decision: they had not chosen a specific animal to kill. Their attack became aimless as some of the nearer gazelles began pronking again. The dancing spread to involve thousands. The hyenas ran into a sea of dancing, darting, zigzagging animals which flowed together into a corporate victim without identity and, therefore, not vulnerable.

Rain clouds fled south. The hyenas became exhausted before any of the gazelles were tired, and they stood together, tongues lolling in the now-hot sun, watching as the gazelles formed up two hundred feet away, the buck at their head again. He divined that the danger was over and dropped his head and began grazing. One by one the gazelle heads went down. The hyenas seemed to understand they had been dismissed and ambled away.

Next morning, the gazelles had gone completely. The baboon scanned all horizons but none remained. He felt their absence; of all the antelopes, only with the gazelles did he have hunting kinship.

The island trembled under the pounding of new rains which struck at midday and continued through most of the night. The flush of grass was so great in the continuing sequence of rain, bright sun and heat, then more rain, that the animals roamed according to no specific pattern; they let the grass lead them on. Instead of gazelles, it was the wildebeests' turn to visit the island. They had completed one of their characteristically wide circles and were returning to the island to be near when the grass recovered from the intense gazelle clipping. The baboon slept fitfully to the sound of their melancholy voices moaning around him.

He watched the wildebeests next day with as little comprehension as he had watched the gazelles. He understood nothing of the complex rituals of territory and hierarchy that were being displayed before him. The problems of territory were more complicated among the wildebeests than among the gazelles. While the buck gazelle had his various alarums and panics, his routines were regular. The female gazelles bred through much of the year, and his establishment of territory did not carry with it the onus of mastering any special group of does. But with the wildebeests, the most exhausting activity was crammed into the few months they spent on the plains. Because their territorial arrangements were more precisely defined, the wildebeest bulls had to keep their herds distinct and their females inviolate during the entire plains experience, though they were forced to move in large compacted numbers. Herds often passed through other herds, yet maintained their identities.

Even before the main body of wildebeests had reached the edge of the plains, many of the bulls were tired. The herds had been almost continually on the move, since the best grazing was seldom close to water, and if it were, the grass was eaten out in a few hours. The wildebeests trekked in long, black, wavering lines which looked from the air like columns of ants snaking across the greening country. In trek, the black bull was calmer than when he was in fixed territory because his cows were usually near him. It mattered little if they were separated from him by a few animals from another herd. There was little competition for the cows during the migration anyway.

Once the herd established itself around the island, however, his supervision became more difficult. He must have his family compacted to his will, its integrity complete, or his work could fail for the rest of the year.

The morning was bright and sparkling with sun glinting through the haze of moisture left by night rains. The baboon contemplated the changing scene around him. The animals moved to what seemed to be self-appointed positions around the island, slowly spreading until some of them were out of sight. The bull wildebeest had no special knowledge of the territory; he had never been to this island country before. He loped around his family and surveyed the conditions. The grass was good; it had grown visibly during the night and would probably outgrow the capacity of the wildebeests to eat it for the next couple of days. When its compulsive spurt of growth had been blunted by their munching mouths, the wildebeests would catch up with it and graze it out.

Two neighboring bulls were too close to the family for comfort, and he charged them. One fled but the other held his ground. The two animals met, bone to bone, and the second bull staggered back. Another, shorter charge sent him fleeing. In the meantime, a young bull had sneaked in behind the family and was trying to drive out one of the young cows who had not bred before. His charge scattered the family. The black bull galloped back to drive off this new intruder, but was faced with a fragmented family, a more agile enemy. His charge sent him into a vacuum.

He faced a new dilemma because his scattered family had run into the territories of two nearby bulls. To recover all its members he would have to encroach into alien terri-tories. He hesitated, then turned to the north and suc-ceeded in running between the resident bull and his family and drove his own group back to home ground. The second section of his family had gone west and was grazing so close to the bull who owned the territory that his new charge took him almost directly at the opponent bull. The attack was diverted by the angry defender. His family ignored his efforts to recover them. He galloped furiously and outflanked the defending bull so that he managed to come on his family and drive them before him. He had suc-ceeded, for the moment, in restoring stability. But during his roundups, the cows and youngsters had eaten well, while he had been unable to take a mouthful of grass. Pant-ing, he dropped his head, but another challenge presented itself immediately. This time, a bull leading his group to new feeding grounds sparked a general movement of ani-mals, causing confusion and concern. The black bull saw a scrimmage of his cows and calves grip the moving family, even as the intruder bull charged. This attack was so vengeful that the black bull's family group fragmented. He saw his harem hopelessly scattered into the ranks of the grazing animals nearby.

He snorted and faced the attacker, unsure whether to fight or go in search of his cows. He chose to fight. He charged and caught the bull at the flank of his own family. His attack dispersed the moving family group. Its members melted into the other wildebeests. The two bulls faced each other alone. They charged repeatedly, senselessly, since there was now nothing for them to fight over. Neighboring animals moved into the disputed territory and grazed it, but that made no difference. The fighting did not end until the intruder bull had been driven off the territorial grounds.

The baboon used the confusion to leave his tree and go hunting. He had looked closely at the leopard's tree and had seen no sign of the cat's dappled form. The hyraxes were out, too, sensing the leopard's absence. The press and movement of wildebeests provided an aura of security. The wildebeest bull spent the rest of the day rounding up his family. He could recognize each individually by scent, but several of his young cows had already been taken by bulls from other families. By late afternoon, when the herd began the trek away from the island toward sleeping grounds in the south, the bull still had not eaten. He had lost two of his older cows in the afternoon melees, but he had accidentally acquired a new one from another bull. That evening, when the wildebeests were strung out in a long line ready for sleep (the best formation for protection against hunters), the wildebeest bull lay among his remaining cows and calves, so hungry and exhausted that he was untouched by the rumbles of lions, the yaps of hyenas, the yelping of distant jackals.

Tensions of territory pervaded the plains and the woodlands, and each of the animals had developed quite different ways of dealing with this powerful drive. In the woodlands, the impala bucks behaved with a grotesquerie unmatched by any other grazer. Each successful buck held sway over twenty to forty does, but to keep control of such large harems, he had to fight off the depredations of bachelor herds while simultaneously responding to the sexual calls of his does. These strenuous activities left him little time for grazing. Some bucks, overly ambitious and too successful in rounding up does, were gaunt for want of food and exhausted from battling rival bucks and from too much sexual activity. In gaining all, they were in danger of losing all, as the bachelors sensed their failing strength.

New grass followed two days of rain and brought a straggle of zebras into the embrace of the island. The zebras grazed in an open group of fifteen thousand animals, creating what seemed to be a contented, conglomerate herd joined in an easy mutuality of grazing. In fact, it was a multitude of family groups, even more tightly dominated by zebra stallions than the cows controlled by wildebeest bulls. The authority of the stallion over his family was rigid. A typical stallion brought his group of fourteen animals to the shores of the island. Five of them were mares and the rest foals. He expected his stallion foals to leave the family as early as one year of age, but a few stayed on with him for two or three years. Some of his mares disappeared from time to time, either abducted by wandering stallions without families of their own or attracted to larger family groups belonging to other stallions. When his mares were in heat, he found it difficult to have total control over their movements, particularly at night when there were many wandering stallions looking for mates. But in the main, he kept his family in a tight unit whose numbers and individuals might not change for months. Now, deep into the plains, all his mares were pregnant. They would be foaling soon, and the births would continue throughout the rains and afterward. This meant special vigilance. Pregnant mares were especially vulnerable to both hyena and hunting-dog attacks, since they were slow in flight. When he had two mares in late pregnancy simultaneously, he could not defend them both without putting himself and his entire family in jeopardy.

The baboon, high in his tree, saw a pack of hunting dogs approaching from the south. They loped along at an easy, unhurried speed, heading directly for the island. Almost no creature was safe from the hunting dogs, although they preferred to attack grazing animals. The baboon had seen fellow baboons cut down and devoured in seconds when they had strayed too far from the main troop. He had watched hunting dogs bait lions and once had seen them catch a leopard. A dozen of them fastened to the frantic, spitting cat and tore him to pieces.

At the same time, the gazelles were moving back into the vicinity of the island from the north. The buck gazelle moved with the group, but he had hardly any regard now for the strictures of territory, so quickly were the animals traveling. They did not trouble to eat methodically, nibbling instead at the choicest grasses. During slow migrations the buck could keep defining his territory by urinating or defecating along its boundaries. When dried grass stalks were common, he marked his territory by thrusting the stalks up inside a pair of holes situated just in front of his eyes. There, a set of glands contained black, tacky material which stuck to the stems of grass. Sometimes, he would mark pieces of grass so often that black blobs of glandular material would bulb on the grass stems.

Now, however, with the movement so rapid and the objective the island again, he did not trouble to mark his territory. He grazed quickly, quite incapable of sensing the approach of the dogs who were then, in the midafternoon, still two miles distant from the island. None of the zebras had seen or sensed the dogs, though their arrival heralded the smell and pang of death.

Although the gazelles' movement was quick, the buck maintained a semblance of territorial integrity merely by his presence. He kept all competing bucks well distant, and as he neared the island, he stopped, looked around, and established that region as his territory. Many of the does kept moving until they reached the shores of the island. The baboon saw them gathering in groups of a score or more in places where trees offered them shelter from the sun. Some of the groups consisted of bachelor males, not yet ready or privileged to possess territory. They were forced to shelter against the island by the presence of the territorial bucks out in the open. This invisible law of territory was so strong that even though the does would later leave the sun-shelters to go out and graze, the bachelors would be held back against the island by the unseen pressure of the territorial bucks. They might remain there until evening, or into the night, when the hunters would come. If the leopard were on the island, he would come on these bachelors first in his night hunting. The bachelors, it had been decided by uncounted long experience, must be the first to die.

The baboon remained fixed in his tree, fixed by the sight of the approaching dogs and by the flight of outlying zebras who wheeled away in groups from the dogs and galloped off. The dogs did not attend them but kept coming ahead steadily, as if the island were their objective. As was their custom, they were strung out in a loose line led by a bitch and followed by the second bitch in the hierarchy, with the submissive dogs and less important bitches trailing behind.

The attack appeared to grow spontaneously, even aimlessly. The lead bitch broke into a faster run which took her toward the zebra group nearest the island. The stallion stood watchfully with his mares and youngsters behind him. Two of his mares were in the last stages of pregnancy and he could not tolerate a long flight from the dogs, but he did not yet know whether his family was the target. The buck gazelle became conscious of the danger, and he ran around the circle of his territory, as if inspecting it, but he did not flee from its confines.

When the attack concentrated on the stallion's group, the entire family turned and fled. They galloped at an easy pace, not in panic, and the dogs strung out behind them were also unhurried in their pursuit. The lead bitch forged ahead, the outrigger dogs widened the front of the attack. The stallion, ears back, saw the dogs appearing on either side of him, saw his heavily pregnant mares only paces ahead of him. He was faced with a difficult decision: to run until the dogs flanked him and then turn to face them, or to swerve now into their tracks. He chose the second course. His swerve was made quickly and was followed by a lightning-fast kick of both feet at full gallop. The kick caught the nearest dog a blow that sent him squealing. The other dogs were almost upon the last of the two pregnant mares, and the stallion swerved again. This time, he could not use his feet and blundered into the dogs who had accelerated to maximum speed. He bowled over the lead bitch. That broke the temper of the attack. The dogs scattered and stopped, panting, while the zebra wheeled, his family compact behind him.

Almost at once the dogs were off again, this time toward the northern side of the island where the buck gazelle held his territory. The sight of the dogs sent the gazelles leaping madly in frantic pronking, but these were not young hyenas, and the buck found himself leaping alone, all the nearby gazelles gone in panicky flight. He heard the pounding feet of the dogs, their eager gasps as they overtook one of the gazelles, but he could not run with the mainstream of flight. Instead, rabbitlike, he took off in a semicircle, running diagonally across the line of attacking dogs. One dog, aiming himself at another victim, made a half-hearted lunge at the buck, but the antelope's diagonal speed was too great. The buck reached the center of his territory, almost exactly the position from which he had started his flight. Although late-moving dogs were still streaming past him, he lay down. The dogs ignored him and in moments dogs and gazelles were out of sight. Only strangled yelps and ripping sounds told the buck how the hunt had ended. Later, other bucks began to arrive back in their territories around him. They, too, slumped down, gasping. They had circled more widely than the buck, but like him, they had been brought back by the pull of territory.

The baboon, a mute prisoner fixed in his tree, hung

there, almost sightless, while the hunting-dog attack turned in his brain. The blood of three gazelle deaths silenced the whistling gasps of air being expelled from the dogs' lungs. The grinding sound of gnawing jaws was clear in his ears. When he focused again, the dogs had gone and so had the dead gazelles. Only a confusion of footmarks and smears of blood on the dark ground remained. The survivor gazelles were grazing as if the attack had never occurred. Night came to the island, but the baboon did not sleep immediately.

VII

Thunder muttered along distant horizons, and the leopard lay limp on his branch, staring slit-eyed at moving gazelles, at grazing wildebeests, at the casual, leisured movements of animals who felt secure. He had no need to turn his head to look into the branches of a tree capping the large rounded rock that formed the top of the island. The baboon would be there, all right, just as inaccessible as ever, and therefore unworthy at this moment of the leopard's attentions.

More than any of the other big cats, the leopard was the individualist, the unpredictable hunter, the silent shadow and night worker whose kill was almost always punctuated by screams of surprise and terror as the victims saw hooked claws coming down at them in the gloom, saw the grinning face emerging from the long grass, heard the crackle of foliage as the big cat dropped from trees. He lived by surprise. Most of his life was spent in the meticulous preparation necessary to create surprise. Unlike the lion or cheetah, or the smaller serval, caracal, and gray cat, he was deliberately self-limiting in his range of hunting, his territory perennially restricted by a major device for survival that set him apart from the other cats and hunters. He was not merely the loner; he was also the most deliberate of the feeders. He preferred to delay his feeding after the kill and to have a larder of meat available at several points along his hunting range. For this, he needed trees, preferably set in good cover, which would provide high storage for his victims, conceal the dead from vultures, and put them above the reach of the greedy, opportunistic lions.

Thus, his territory tended to be linear, a stretched-out range that might wind for miles along a watercourse. It was the nesting place for cranes, the home of monkey and baboon, of dainty bushbuck and ground-prowling birds like the francolin and guinea fowl, who all used this sheltered territory to avoid the special dangers of life in the open. As an individualist, he was as remote from group ethic as it was possible to be, and he possessed special traits which made him unexpectedly dangerous to creatures he did not actually eat. He hid his prey, but vultures did find the meat occasionally. The ferocity of his attacks on them when they were bloated with food or vulnerable in some exposed night roosting place, sometimes led him in vengeance to kill three or four of them at once and leave their bodies to be eaten by other scavengers. This enmity pushed him to silent daylight attacks on vulture nests set in trees. There, he smashed eggs, hooked youngsters out and let them spin away to earth.

But his greatest rages were provoked by hyenas, those omnipresent competitors and followers who always seemed to know where he was and what he was doing, and who were capable of working together to deprive him of his victims. Their squealing, yapping, moaning voices were maddening. Their mob tactics, rushing charges, and sudden retreats, endlessly repeated, invariably unnerved him if he were on a kill so that he fled, body low and streamlined, his rage intense as fire.

The cool afternoon warmed slightly when sun watered through gray western clouds. The leopard dozed. Hyenas did not always travel in packs, and he was often tempted to hunt by day when it was more likely for solitary hyenas to be abroad. The rains had brought hyenas down out of the northern woodlands, and they often traveled alone. As he watched from his tree, an old female, heavy with milk and obviously hunting far from her den of cubs, appeared on the northern horizon. Her round-eared, slovenly face came into focus, and the leopard slid down his tree, disappearing into the grass. The sight of the cat coming down the tree stopped the hyena, her memories of leopard present and sharp. She could punish her lumbering body to fifty miles an hour in a burst, but this was little use against the leopard's sixty-mile-an-hour attack. If she did not flee at once, she might die.

The hyena could not now see her enemy hidden in the grass. Her blind eye narrowed her vision. But the leopard saw his victim clearly, every movement of his attack already sharp in his head. The satisfaction of the kill would be special, vengeful really, because he would not eat hyena flesh. He had once mastered a territory flanking both woodlands and plains, and he sometimes had as many as six hyenas hanging from the crotches of different trees spread half a night's walk down the winding route of a watercourse treeline. The disintegration of these bodies took many forms. They disappeared quickly under the wrenching beaks of vultures who found them; slowly under the columns of ants who sensed meat high above them; relentlessly under the wriggling masses of fly larvae who consumed totally any body not found first by larger scavengers. The female hyena turned away, caution guiding her, and took herself beyond the range of the leopard's charge.

At dusk, he was on his branch one second, gone the next. He dematerialized so quickly that the baboon, watching, blinked at the disappearance. No living thing saw the leopard's passage down the slopes of rock, between the defile of boulders, among the clustering shrubs near the shore of the island as he headed toward the plains. This was the perfect time to leave; the moon not risen and the possibility of darkness all night with the threat of rain. A quail shot from under his feet in an urgent slash of wings and disappeared. The aroma of hare hung thick across his nostrils for a second and he paused, a favorite food this, but the scent passed on abruptly, and he understood that the night wind had collected it from afar, compressed it into one seductive message, then whisked it away, giving no hint of the hare's location. The leopard walked on. This early-evening moment soothed his appetite, blunted his sharp awareness of the need to kill because none of the possibilities of the hunt had been tested and so there had been no disappointments.

Instead, images clicked behind his smouldering eyes as he shuffled and sorted what he knew about the apparently blank darkness stretching ahead of him.

He never nightwalked without such computations. The smells of the day might come to him from a great distance, but coupled with the memory of other nights, other smells, they gave him a prescient knowledge of exactly who sheltered in every patch of long grass or at the top of which tree; who had passed along which pathway, and when. He walked forward but never blindly, never without the lethal awareness that made his victims so helpless in those last moments of the hunt. Now, he stopped. One of his favorite victims lay north of him, fairly close, the smell of aardvark so strong that the leopard was cautioned to pause before considering the attack.

Caution dwelt in the memory; these clumsy, harmless-looking creatures could frequently surprise the hunter. The leopard's technique for killing aardvarks was the same as that which he used for dealing with porcupines—one lethally fast blow that disemboweled, crippled, or killed before the victim knew the leopard was near. But time and again, he had lost aardvarks, even after grievously wounding them, when they had ripped free from his grasp and ducked into a nearby burrow, any burrow, to disappear with puzzling speed.

The aardvarks were moving across the plains this season in response to the flush of termites everywhere. They dug into deeply excavated earth, their long tongues coated with adhesive mucus and flicking forward into the tensely struggling termites. The leopard understood the wariness of the aardvarks and their sensitivity to danger. Flattened, he slid north.

The aardvark had just moved into position beside a mound of freshly dug earth which indicated a large and healthy termite colony. For the leopard, the best time to attack was just at that moment when the aardvark was preoccupied with his own excavation and the prospect of eating. The leopard crouched, and waited.

The speed with which the aardvark penetrated the termite city was precisely calculated, since the termites themselves reacted swiftly to any outside disturbance by jamming down galleries and corridors into deeper earth positions. Therefore, to get his meal, the aardvark had to dig into the city and disrupt its network of galleries quickly so the flight of the termites would be blocked. Already, he could hear the panic of termites as they responded to the vibrations of his footsteps. He plunged forward, digging furiously.

The moment the aardvark's stumpy tail was the only part of him showing out of the ground as it flicked back and forth to kick aside the dug earth, the leopard charged. He got both paws deep into the hole beside the aardvark's body, dug in his claws, and with a tremendous jerk pulled the creature clean out of the hole. But the effort needed was unexpectedly hard and he stumbled at the moment when he should have killed the creature. Instantly, the victim curled himself almost double, becoming nearly spherical in shape to protect his soft underbelly. He bent his vulnerable head and long nose out of the way, so that the leopard was forced to grapple with his immensely strong and curved back, the thick hide of which was already scarred by the marks of other attacks.

It was difficult to get a grip on the back. The skin, tough as leather, resisted the leopard's claws. His teeth did not penetrate it. He grappled, clumsy and growling, while the back legs and tail of the victim flailed with convulsive movements so powerful that the cat was finally flung clear. The aardvark uncoiled, leaped for the hole in the termite city, and sank into the ground in a frenzy of digging. The leopard pulled him out again, and was flung off so strongly he was partially winded. When he recovered, the tail of his victim had almost disappeared down the hole.

Making a grand effort, the leopard dug his claws fully into the flanks of the animal, but this time, he did not have the strength to pull the creature out. While he strained, cat body bowed with a sustained effort that was alien to his personality, he could feel the aardvark trembling with the furious energy of digging, could feel the earth being flung up into his muzzle, the tail flailing at his fangs. He got a grip on the tail but by now his head was so deeply into the hole that he was having difficulty getting purchase with his back legs against the lip of the burrow. He felt himself being drawn down into the hole by the strength of the aardvark. Finally, disgusted, he let go. Coughing and sneezing, he backed out into the cool quiet of night. Far below, he heard scuffling noises as the aardvark dug deeply into the earth. Then silence. He licked his paws and chest, rubbed his face with each paw, and held his rage and frustration in the center of his body.

Like all cats, the leopard was accustomed to feeding spasmodically, sometimes eating no major meals for several days, then gorging on a big kill. The failure of the aardvark hunt would have meant little in the normal course of his life, but he was not in a normal situation. His present hunting territory was impossibly elongated, the distance between trees and between the smaller islands far too great.

He had gone north to the permanent water hole where fine trees grew and a dozen small islands gave good cover, but this ground was denied him by another leopard who held dominance over the four-mile-long territory. The leopard could hunt the fringes of the water hole briefly, as had been his custom, but he was prevented from pressing through and beyond it to the woodlands by the presence of the resident leopard.

So, each night was a distillation of tension and frustration that made the leopard irritated and savage. He was continually on the point of being aggravated into a desperate act which would either resolve his dilemma or kill him. Now, his toilet completed, he contemplated the night again. The soft whooping cries of a pair of hyenas slid eastward. A guttural cough warned him that the other leopard was in his territory and would defend it, but the direction of his voice betrayed his whereabouts; he was at the northern end of the territory. In the trees on the nearest island of the water hole guinea fowls had made a new roost, this information remembered from previous nights. Now, with failure recent, the leopard became haunted by the flavor of guinea fowl flesh, the most succulent of all kills. Not even mouthsful of feathers could diminish the delight of such soft red meat; not even the brevity of his satisfaction—the meat digested so fast—could change his interest in where the guinea fowls were roosting. In typical cat fashion, he spent an hour in what appeared to be aimless wandering, criss-crossing from scent to scent, from empty burrow to fresh dig mark, before coming to the island.

It was less an island than a sprawl of rocks, thickly treed and protruding only slightly above the level of the plains. Its western end petered out in a scattering of scrub and shrubs which almost reached the shores of a second, larger island. The islands were arranged in a natural amphitheater around the water hole from which bled the small flow of water that sustained a meandering of trees leading away to the southwest. The silent walk of the leopard did not alarm the sleeping birds. He leaped upward three times the length of his spotted body, and did not waken a single bird, his body weight exactly balanced so that there was no scuffing of paws, no shaking of the tree at the end of the leap. He seemed to have been transformed from a terrestrial to an arboreal creature with no hint that any muscular effort had been involved. The night air breathed a soft wind, and tiny pinpoints of sound brushed off the sharp tips of acacia thorns.

The moon appeared in a cloud rift and revealed the delicate curl of the leopard's tail against its face. The guinea fowls were roosting high and remained asleep in the illusory security of height. A second leap, more daring, took the leopard to a higher, thinner branch, and this time, the tree trembled. The birds awoke, quick necks darting tiny-brained heads in every direction, concerned eyes running the length of the leopard's pelt in the moonlight, but seeing nothing. The moon shone through grasping foliage, its light fragmented by its passage, and fell on the dappled back of the cat in a confusion of spots too difficult for the simple brain of a guinea fowl to comprehend. Their concern gradually diminished; the trembling of the tree slipped from their memories.

The leopard's awareness was almost total, and he gave attention to every detail of his situation. Now, he heard a small movement and stared down. Below, he saw the patient watcher, the deferential hunter, the silent, skulking figure at the edge of the crowd, the master of the quick run to seize stray scraps of meat—the jackal whose anonymity could make him dangerous to those who dismissed him as a threat.

The jackal had earned his unique place in this day-and-night world without special talents or skills. He had no lethal bite, no rapid surge of speed, no comfort of pack numbers, no great strength, no exceptional acuity of vision (he did not now see the leopard), no unusual capacity to trail by scent (he could not smell the watching cat), no well-refined powers of communication. Instead, his skill was his capacity to connive, to wait, to be the sly presence standing at the hocks of the hyena, the attendant at a feast of carrion-eating storks. He was a daytime fox in the raiment of a small wolf, a dog with the mannerisms of an unfinished cat.

If he paused for one trembling second now, and looked at the patch of moonlight ahead, it was because he had that special intuitive sense of danger possessed by those creatures caught in an evolutionary trap between hunters and victims. A meat eater, he was also eaten. He fell under the hyena's massive bite and was eaten with relish. Hopeless, he fled the cheetah's blinding charge. Worst of all, he was the plaything of lions, often caught in one of his unwary rushes at the edge of lion kills, and turned into a wretched toy for the young lions. He was tossed into the air, thrown from one animal to another, sometimes crippled, sometimes killed, but just as often let go when the lions got bored. Then, he would drag himself away, horribly mutilated, while the cats went limp in sleep. So he looked at the moonlit patch, trees surrounding it, and comprehended danger.

The flesh of the jackal attracted the leopard, not so much

for its flavor but because it represented a full meal, whereas a guinea fowl was only a morsel. Even two guinea fowls did not make a true meal. The guinea fowls, rotund featherballs sustained in ignorance, shook themselves and turned to sleep. The jackal, though, could not be a sure target. A branch bisected the leopard's view of him and obstructed any chance of a free-falling attack. He could spring outward, but this would surely rouse the guinea fowls. Indecisive, the leopard waited.

In the long silence, punctuated only by the hollow, booming cry of a distant eagle-owl, a new scent carried on the north wind came to the leopard. It was a strong one, familiar, and almost as attractive as the smell of monkey or baboon. Soon after the scent came the sound of the animal's approach, a careless clattering that immediately wakened the guinea fowls again. If the porcupine were careless, it was an attitude born of the special confidence given him by his defenses. A waddling bag of delicious meat, the porcupine's tender flesh was the antithesis of his sharp spines. His bravado by daylight took him under the nostrils of lions who, trembling, controlled themselves if they had already suffered spines in the muzzles or paws. If they had not, they often made foolhardy attempts to penetrate the porcupine's defenses. Paws swung, the porcupine rolled, its spines flying like lethal arrows. The attack was followed by the anguished roars of lions who faced lame days and nights that often ended in death by blood poisoning or starvation.

Only the leopard had made an adjustment to the porcupine's defense. He killed the armored bag of flesh with a vicious, underhanded swipe of claws that caught the animal on his vulnerable flank where the spines pointed rearward and lay flat against the body. If well placed, the blow

132

half-eviscerated the porcupine. The spines might rise in hackled and hopeless recognition of danger, but too late to avert death. The leopard would wait until the death agonies were over before cleaning out the body from inside the bag of quills, leaving them as a twisted relic of his success.

In long grass, with bushes nearby and trees overhanging his route, the porcupine, spines clicking, grunted along in almost total safety. He saw the jackal and ignored it, even making it step aside hurriedly as he walked, undeviating and arrogant in his assumed security. The leopard had the porcupine in clear view; he noted the relaxed position of the spines and salivated at the prospect of such sweet flesh in his mouth. But an attack meant an accurate, soundless free fall, the blow of his paw just fractionally preceding the landing of the other three paws. To tempt the cat further, the porcupine stopped in clear view with moonlight touching him as his pointed snout lifted in quest of invisible food ahead.

At that moment, a guinea fowl began a disbelieving cackle of horror as her eyes transmitted the outline, but not the design, of the leopard's body silhouetted against the single patch of moonlit ground. It was a cackle the leopard understood, and it snapped his attention away from the porcupine. Imprinted in his memory were the roosting positions of the guinea fowls and he had no need to reconsider them. Even as the cackling bird's voice rose to a scream that blossomed from the treetops, the leopard's body tensed, twin paws outstretched for two victims in the gloom above. The jackal flung himself away. The top of the tree exploded with birds thrashing blindly to anywhere.

The uproar in the guinea fowls' tree was transmitted instantly to every other part of the complex of islands. It was heard by large owls who hunted guinea fowls; by three hyenas who sometimes caught guinea fowls crippled by more agile hunters, and by jackals, small cats, and a civet. It aroused sleeping vultures, who understood what the noise meant. It caught the attention of the resident leopard, who divined more accurately than any of the other listeners what had happened. Before the intruding cat could get both guinea fowls jammed satisfactorily into his mouth, the resident leopard was into the tree beneath him. A throaty, rumbling growl of liquid hatred rose to the top of the tree.

All the animals on the plains and in the woodlands were linked by the discomfort and hostility they felt when they were hunting in the territory of others, or when others were hunting in their territory, but territory could only be enforced within each family of animals. The lions could not drive out the leopards, no matter how much they hated their presence. The leopards could not discipline the cheetahs—who had no territory and wandered free throughout the year—except to steal their victims when they killed close to cover. The leopards hated the lions but remained powerless to influence them.

What made territorial interaction possible was that each of the hunters had different skills. Sometimes they overlapped, and that caused conflict. The leopard might have an appetite for ground birds, but his speed was nothing compared with that of the caracal, who could kill three or four francolins before they were able to get their wings working, kill another two as they got into the air, then pull down two more with one last great leap. For sheer ferocity, the leopard could not equal the gray cat, a creature not much bigger than one of his paws. The leopard killed caracals or servals if he was lucky enough to surprise them, but in his one encounter with a gray cat, his whole muzzle and both paws had been so lacerated by the vicious response of the small cat that he was wary of a repeat encounter.

Now, he felt the supreme discomfort of being in alien country with its owner beneath him. He was torn between the need to fight and the fact that his mouth was full of food. His disadvantage was that he did not belong here, and he knew this. To contest the cat below, he must have an overriding impulse which he did not yet possess. The reaction to the challenge was still ambiguous, a combination of the need to survive and to preserve his food while simultaneously testing the determination of the territorial cat. With a deft motion of his head, he wedged both guinea fowls into a crotch of split branches and prepared himself to fight.

To be trapped in the top of a tree: no leopard could stand such mockery of his freedom. But to be trapped there by another leopard was unbearable. Better, perhaps, if he were to relive the frequent times he had been treed by lions. The previous year, he had caught a gazelle in bright daylight, seizing the animal in a quick charge from cover, but knowing, as he vaulted into his tree, that he had been seen by lions. One of them yawned and got to its feet, even before the leopard had wedged the gazelle into place as high as he could lift it. The lion moved toward him and the others followed. He had two options open to him: he could flee, or he could wait and actively defend his food. The lion could never climb as high as the leopard, but lions generally were so clumsy in trees that the leopard could not remain calm while watching one of them beneath him, scrabbling at the trunk, hooking great paws into inadequate branches, ripping them off as it half fell, half hung on. Nevertheless, this time the leopard chose to defend his food. The lion climbed awkwardly, and the entire tree shook back and forth as though in a storm, finally dislodging the wedged carcass of the leopard's kill. He was left with the double mortification of losing his kill and of still being treed. The lions gathered around and ate beneath him. As the final indignity, they went to sleep, sprawled out so widely that to reach ground he would have had to jump almost on top of one of them.

An owl cried again; a chittering of hyena giggles erupted in distant darkness. The leopard dropped silently to meet his enemy. Both cats faced each other on the same branch, bodies twisted, tails lashing, muzzles almost touching as they growled and hissed, the air around them made venomous by their hatred. Because some leopard fights were to the death, the rituals leading up to the fight were often elaborate. Here, however, with no room to move backward, forward, or sideways, the two cats were forced to act quickly.

In an explosion of movement, both animals fell off the branch. They hit the ground fighting and screaming, limbs blurred in rip and counterrip, back legs raking, claws reaching for eyes and underbellies, teeth searching for strangling grips. Their bodies, contorted into impossible shapes, rolled away from the tree into the bushes. Everywhere in and around the islands animals paused, or wakened, and listened to the sounds of the leopard fight. This was more truly catlike than the thundering of lions. The vicious screams and yowls, the sounds of ripping flesh, the crashing and splintering of undergrowth sounded like a dozen large animals in combat.

As quickly as it had begun, the fight ended. The leopard was straining to get his feet up into the resident cat's belly for one tremendous rake that would cut it open, when his grip slipped and the cat leaped away. He knew he had hurt the other leopard. One bite into soft flesh had brought a gush of blood, and deep inside his head he had heard a bone break. Now, tail lashing, he awaited the next move, but nothing happened. The other leopard had gone. The bushes crackled and snapped as they tried to straighten themselves from the crushing of the cats. Silence. The leopard's apprehension of the unknown and the uncertain increased. He had not been so badly mangled that he had been forced to flee. But neither had he thrashed the other cat into submission, forcing its retreat. Either of these solutions would have resolved his conflict of territory. Flight would have sent him straight back to his island. Victory would have ensured this new territory for himself. But no firm answer came from the night air, darkening now under a mask of black clouds.

Without clear orders, he resolved the tension by returning to the tree, lightly climbing it and recovering the two guinea fowls. Then, head held high, he walked slowly west between gray rocks toward the water hole that lay in the center of the island complex. He had taken a dozen paces when he was struck in the back of the neck by a blow so powerful the guinea fowls exploded from his mouth. He yowled and rolled, another body atop him, but this time he was so defensive he had elected not to fight even before the fight had truly begun. He kicked himself free, half saw the other leopard's face close to his, and then he was away at full speed, hurling himself among the rocks and heading, willy-nilly, for the open plains again.

The worst nightmare for any leopard was to be caught in the open, in daylight. Once, in foolhardy pursuit of a bushbuck he had flushed from cover, the leopard had found himself hundreds of feet away from his beloved trees by the time he had made his kill. His presence in the open had been marked by baboons, his traditional night victims, and they had sought revenge. Forty or fifty of them had dropped from the trees and begun to spread out, loping toward him. No single baboon could hope to match his strength, but a troop so big could easily kill him. With the baboons converging on him, he had been indecisive for a moment, vacillating between running away or boldly charging them. He had elected to charge, but too late, and a dozen of the big-toothed creatures had fastened themselves to his flanks like obscene leeches, their teeth ripping his hide to shreds. He had rolled and swiped and shaken them off, killing two of them and then had driven mercilessly at the others, scattering them.

So now, racing away from the security of the water hole, he looked into an eastern sky already lightening and felt apprehension. He was far from his refuge island. Though he had no need to fear baboons here, there was every chance he would meet other group-hunting animals before he reached the island. Like all cats, he had enormous strength, but little stamina, and his run quickly slowed to a trot, then to a walk. His wounds bled. He could look down his nose and see the jagged stumps of his whiskers splayed out untidily where they had been chewed off.

A dark thread of creatures appeared in the smoky light—wildebeests stretched out in a long, compact line of sleep. He could not flank them. They did not see him until he was close, and then they erupted in panic flight, splitting into two herds which roared away together. Chill, predawn wind touched the leopard's eyes, and the sun appeared, blazed flaming red behind the wind, and he was running again, directly into it.

Because the rains had stimulated vast movements of animals, the leopard had no way of knowing what creatures might be in his tracks ahead. There could be tribes of hyenas, or nomadic lions, or worst of all, hunting dogs. The dogs were rangers, unbounded by territory, and capable of moving hundreds of miles in a month. They traveled from one end of the plains to the other, and there was no predicting where they would be at any time.

But it was not the unpredictability of the dogs that affected the leopard; it was the ferocity they felt for him. Twice in his life he had been intercepted by dogs while he was in the open; once, he had been only a score of bounds from a tree, but both times he had nearly died. Now, a sickening fear ran through his body as he saw the dogs. They had not yet seen him and were still in a tight family group. Some of them were gamboling in play, licking each other's muzzles, preparing themselves for their usual morning hunt. The leopard had never seen so many dogs, far beyond his capacity to count. He flattened himself as close to the ground as he could get, his elbowed legs working like pistons as he strove to run in a stalking position. But he was seen almost immediately. He made a slight change of direction and the island appeared in silhouette against the fast-rising sun. The dogs stood bathed in bright pink light, their blotched hides melting into a multiheaded animal that watched him.

The decision to attack or not was made by the dogs' leader. She turned away from the leopard, toward the island, in what was intended to be a flanking movement that would cut off his flight. The rest of the dogs strung out behind her, and the despairing leopard saw that they would interpose themselves between him and the island within minutes.

He could easily outrun any dog in the sprint, but he was tired and wounded, and he knew intuitively that the last of his strength would be needed for the final rush to the island, if he even got that chance. The dogs were flanking him easily, loping along in steady, distance-consuming strides. The leopard realized he could not now outrun the dogs and chose another tactic. He slowed and bore slightly right to allow the dogs to get well ahead of him. They did not understand the strategy and outflanked him quickly before the leader saw her mistake and brought the group to a halt. Here was the baboon dilemma repeated. The leopard mustered full strength and charged.

The working strategy of the hunting dogs was to come up behind their victim and pull him down at a full gallop. They could do this easily with the leopard, but they were less well adjusted to the notion of an assumed victim running full speed *toward* them. For their time-tested plan to work, they would have to let him pass them and then pull him down on the run. Before their indecision was resolved the leopard hit them.

He got two dogs as he crashed into the group, mortally wounding them in eviscerating swipes that sent intestines and blood flying. The screams and moans of his victims mingled with the choking gasps of dogs striving to leap away from him and dogs straining to close on his speeding body. He got a leg in his mouth and snapped it off like a dried stick. The leg ripped from the body with a snap of breaking tendon and he dropped it instantly. Quickly and easily, he outran the dogs, his lean, bounding body eating up the distance to safety.

140

But his magnificent cat muscles were instruments of the briefly impossible; they could not match the predictable, unexciting steadiness of the dogs. He was not halfway to the island before they began to overtake him. They never changed their running tempo; that was their strength, the ordinary made lethal by extension. The leopard's body was supercharged with pain. His limbs, his lungs, his gut cried agony. He saw the island through a crimson curtain as he felt dying vestiges of energy pumping into his muscles.

Dog and leopard reached the shores of the island together. One dog got a grip on the cat's tail but, weak-jawed, could not hold it. From another miracle of the leopard's muscles came a giant leap which left the dog with a tuft of fur in his teeth. The leopard's jump seemed an act of easy, powerful grace, but when he got into the nearest tree and stood on a limb, he was the lowest and most tortured kind of refugee. Gasping air into his lungs, he recovered while the early sun warmed his ears and the dogs sat on their haunches baying at him. Their voices contained a warning of the inevitability of time. The leopard might escape this day, they seemed to say, but there would be other times, and there would always be other dogs.

VIII

The baboon saw the return of the leopard and listened apprehensively as the baying dogs milled along the shores of the island. But they were in alien territory so close to its rocks and sheltered places. Another leopard might be concealed in there, perhaps a female with cubs. They did not stay for long. When the dogs had gone, the baboon watched the leopard slump down awkwardly, his rump wedged in the split trunk of the tree, and go to sleep immediately. The baboon sensed the tension in the cat, sensed he was beset and sick since he had made no effort to survey the island. The baboon was emboldened by the misery of the leopard to leave his tree. He quickly let his hunched black body down the side opposite the leopard and ran rapidly among the rocks, sending hyraxes scampering out of his path.

The appearance of the termites floating past the island, the sight of them pouring out of their mounds, had aroused an overpowering appetite for them, and the baboon remembered that a particularly large termite mound to the south had been swarming with creatures when he had hurried past it late one afternoon. He scampered down the last slope between rounded rocks and ran out onto the plains. The distant forms of gazelle and wildebeest, of zebra and hovering vulture faded as he entered the real world of the plains, the world where the decisions of life were arbitrated with an intensity and an absoluteness that had no match in the more sheltered woodlands. There, his sharp eyes flickered intently as they saw the endless bounty of insects which jumped and crawled and flowed in and out of the ground.

In this microscopic world, the baboon became an insignificance of matter compared with the billions of creatures inhabiting the earth beneath his feet. When a sudden rush secured him an extremely large dung beetle, he did not notice that he had dragged one foot through a column of marching ants until he felt stings on the sensitive skin around his ankle. He did not look back, indeed had no interest in ants of any kind unless some accident opened up one of their nests and exposed thousands of eggs or cocoons. Although he might try then to make a meal of some of the cocoons, he did not relish being stung so severely by ants who fought back viciously; the meal simply was not worth it.

He was diminished now because he, animal alone in this place, produced no measurable effect on the world around him, whereas the ants, in their mysterious millions, helped to control insect populations whose activities determined the quality of plains life. The baboon had placed one foot fairly in the middle of the densely packed ants and flattened all the creatures beneath. His scuffling second foot completed the break in the column, sweeping the crippled and the dying away from the line of march, smashing the elaborate ramparts of damp earth which the ants had flung up on either side of their route.

The baboon's blundering feet upset a critical, complex, and delicate organization. At both ends of the break milling clots of excited safari ants gathered. They were only the distance of the baboon's long foot apart; yet they were unable to make contact with each other because they did not move by sight or by the common direction of the road. They navigated by scent, and this invisible link had been broken at a critical time, just as the safaris had reached their prey in the forward attack zone and the hunting was yielding its first victims. According to the strict rules of the safaris, these victims must be transported back to the bivouac area immediately for processing. Any break in the column of march at this moment dammed up impatient hordes of creatures pushing to the attack and stopped the returning booty carriers, making them vulnerable to aerial enemies.

Once the attack had begun, there was no way to stop it. The assault group thrust forward, creating its own kind of microcosmic chaos. Panic riffled away from them on all sides as their jaws ripped into wriggling caterpillar bodies, seized small beetles, pulled away eggs fixed to blades of grass, and grabbed for the main prize, a concentration of cockroaches in a small patch of longer grass somehow left uneaten.

Neither time nor effort was wasted in this smoothly organized operation. As fast as each victim was killed, he was dismembered and the pieces distributed among the carriers. Nothing was lost. Legs were ripped off, wings severed, body carapaces torn to shreds, each fragment exactly the correct size for rapid transport back to headquarters. This was the internal picture of the attack, most of it hidden in the green thicket of stems. Quite another view was visible above the grass.

The panicky insects, both the sought and the unsought, fled. For those without wings, the aim was to reach as high into the grass forest as possible. Frantic cockroaches and caterpillars swayed at the tips of bending stems while relentless ants climbed beneath them. Moths hurled themselves into the air. Grasshoppers flipped end over end in wild escape jumps. The disarray of the insect hordes, however, was only a small part of the total impact of the safaris' advance.

Several score of small flies moved ahead of the safaris, placing themselves at a uniform distance from the advancing attack zone. The moment early moving cockroaches appeared, the flies dropped down and fastened themselves to the heedless, hurrying bodies, then arced away. In each attack, they left behind them a singly placed egg. As long as the safaris did not nullify their work by killing the host creatures, these eggs would hatch and the larvae grow to half maturity inside the bodies of the cockroaches, eventually killing them by eating them from within.

The predatory flies thickened in their busy exploitation

of the slaughter, but they were soon only a small part of the aerial attackers gathering along the line of the safaris' march. The flycatchers were first. They scattered to get the best share of the disruption below, and fleeing moths were snapped up instantly. Some finches passing overhead saw the feast and wheeled to make a landing, opportunistic harvesters of the safaris' victims. The birds were absorbed by the commonality of such a shared opportunity, so well absorbed, indeed, that they frequently became victims of larger hunters who had observed their preoccupation. Peregrine falcons, familiar with these scenes, knew where to wait and watch, knew exactly which bird to kill first and what mode of attack to use. But for the moment, no peregrine appeared and a slow-hovering, distant hawk did not move closer.

Meanwhile, the break in the column had not yet been repaired and the black turmoil of booty-bearing ants had coalesced in a thick mass at one edge of the break. Small ants, deprived of the discipline and order of the marching column, darted aimlessly about. Some ants met and touched antennae across the barrier of the footprint, but the transfer of scent was not yet strong enough to bridge the rupture.

A francolin cast a long shadow over the ants. His head darted down again and again, and the ants were flung into further disorder as the bird selected his booty, nipped it up, and shook clinging ants from it before swallowing. He worked quickly, feeling attackers moving up his legs. Unless he protected himself, these determined specks of life would eventually reach his thighs under his feathers and there would delay their stings until they produced the best effect. The stings he would feel for days afterward. As the first ant crawled under his feathers, he stepped away from the column and preened himself clean of attackers. But the disruption he left behind seemed total. The column was scattered so badly at the supply end that it was nothing more than a ragged front of confused creatures who appeared to be incapable of restoring the order that had so beautifully animated them scant minutes before.

However, no catastrophe that occurred along the route of march could be transmitted to the attacking forces, and indeed, no useful purpose would be gained from their knowing what was happening at the rear. The kill was scheduled to continue until all life in the attack zone was either wiped out or driven off. The ants butchered their victims and seemed oblivious to the rush of air made by birds' wings above their heads, to the smack of beaks closing on captured insects. They had to work without protection, but behind them guards flanked the hurrying column of insects and stood so thickly in places with their pincers raised menacingly that many ants were unable to force between them. The guards were so sensitive to danger that when a hunting bird swung low, they lunged upward blindly in a response that was at once intuitive and savage.

But there were never enough guards to protect every inch of the column, every branching roadway that diverged from the main highway of moving insects, and it was at these places that small spiders lurked, waiting in ambush for their chance to rob the ants. The spiders hid until they saw a morsel of food small enough to seize, then they jumped with all the speed and bounce of small cats, grabbed the food, and just as quickly jumped away again.

Their raids were so fast that the ant line milled for a second at the attack point while the furious ant who had been robbed sought to find out what had happened to him.

Other spiders had preyed on the safaris for so many millennia that they had changed their bodies to look like ants. The disguise was so successful that a spider hunter, coming across one of them, would pass on, assuming he had seen an ant. Some spiders had so perfected their changed appearance that they were indistinguishable from ants, even to the ants themselves. Thus, with a boldness that could hardly be matched by any other individual, they joined the hostile, marching ants, moving freely over great distances without being molested or recognized by their enemies, until they seized food and escaped.

Despite their fallibility, the safaris overcame difficulties through the sheer press of their numbers, three million of them backed up behind the attack force. Thousands of them could search and fail to find an objective, while it only took one to ensure success for millions behind.

Through hundreds of random contacts, the safaris gradually built up a reconnecting link between the two split ends of the march. The guard ants quickly positioned themselves, while engineer ants rebuilt the roadway and so unleashed the flood of transporters and the rush of new creatures pouring toward the front lines. At the same time, other engineer ants were building a subterranean highway designed with blind skill and energy to carry the line of march under, and out of reach of, the disaster that had struck the ants.

These engineers labored with such haste, however, that in places they worked too close to the surface, and the road above fell in on top of them. The booty carriers, the soldiers and the reinforcements tumbled into the holes, filling them in seconds as other ants piled heedlessly over the top of them. The line of march bumped and twisted to the surging of ants. They struggled out of the holes, shoving aside those on top of them, while the engineers tried frantically to repair the damage. Out of this chaos, incredibly, came order. The safaris separated into twin groups, holes opened up alongside the upper road, and half the stream of booty carriers and reinforcements ducked underground to take the second highway. With all communications reestablished, the ants threw the full weight of their numbers into cleaning up the attack area.

When every living creature had been dealt with, the laden lines of returning ants found their passage unimpeded. Somehow, the information had been sent back to headquarters that no more reinforcements were needed. The hunt was over. The birds scattered. The ants slipped back into the earth that had so mysteriously flung them up into the harsh light of the African sun.

Meanwhile, the baboon had long since finished with his beetle meal and was working his way steadily through the grass, picking at an unexpected source of food. Scattered everywhere in front of him were small, dun-colored moths, so regularly placed that he could be almost certain of finding another one a hand's length away from the one he had just picked up. This was typical feeding for him, like the nipping of the green stems of grasses or the patient filling of the gut with uncounted thousands of seeds, shoots, eggs, insects.

The moths had come to the plains and the island while the baboon slept. They had traveled unnoticed except by night-hunting birds, and their numbers were beyond measurement. Their arrival at this place was part of a general advance which was spread across a front scores of miles wide and several hundred miles deep, with wave after wave moving steadily northeast. Rain had stimulated their movement, but mystery shrouded the reason for such a massive migration. Why should it occur this year and not the previous year, or the year before that? Like locusts, their time had come, and they obeyed the imperative to move north.

The baboon ate steadily among the creatures who had landed on earth softened by showers and begun to smother plant stems and leaves with between a hundred and seven hundred eggs from each moth. He ate among a relatively small number of moths—forty to fifty million landed in a broad band north of the island—but with the air hot, moist, and clear, the hatching of their billions of progeny would take only a couple of days. Such a large number of caterpillars would be capable of producing a recognizable effect on the low, sweet grasses, eating large patches down to earth level.

When they had transformed themselves into moths, their advance would acquire the force of a hammer blow at the landscape ahead. Like locusts, the moths functioned independently of their individual success or failure. They were group creatures, a conglomerate which became annihilatory once growth started, a force so pervasive that only self-destruction seemed able to halt them. Their presence around the baboon was only a tiny part of the total migration, which would be measured by the speed with which all the eggs could be laid down and hatched; by the rapidity with which the caterpillars could stuff themselves, pupate, and be transformed into another advancing army of moths. If all went smoothly, the grass around the baboon would be eaten down to its rootstalks, nothing left even for the gazelles, and the caterpillars would be pupated and off again within sixteen days. Somehow, the growth of the caterpillars and the development of the pupae would be synchronized so that the latest-laid eggs would produce faster-growing caterpillars or faster-pupating moths. Whatever the secret, the moths would leave en masse. They would make an eerie sight, billions of creatures strung out along the great width and length of their migration, white hind wings fluttering in flight. Far beyond the baboon's ken, this irresistible army would alight and deposit there an unimaginable number of eggs.

The hatching of these new caterpillars would become an event beyond reason. Hundreds of frantic caterpillars would compete for food in every square foot of earth. Grazing animals would be unable to feed because of the mass of them. The grasses would be eaten, fouled, consumed to earth. The trees would be eaten to the branches and shrubs stripped of foliage. The earth would be desolate to every horizon of infestation. Small animals and even antelopes would starve among the caterpillar army, even as they mashed them underfoot with every step they made in search of grass that had disappeared.

The contented baboon ate on, delicate forefinger and thumb lifting moth after moth into his mouth. He munched, scanned the horizon, munched, while in the vacuum of his ignorance the muffled thunder of an unborn army sounded.

The moth migrations had to end somewhere because no landscape on earth could continue to withstand these invasions. Thus, some generations of migrants were fallible, unaccountably not producing eggs or being wiped out by unexpected collisions with hosts of enemies. The baboon, ignorant and uncaring, had already been involved in one such collision when his troop had come upon a caterpillar army busily destroying the long grasses around a water hole many miles to the north.

At first, the baboons had been preoccupied with trying to hunt the smaller creatures who were hunting the caterpillars, and for a while it had been a feast of mice and frogs and lizards and other insects. Despite the appearance of hundreds of marabou storks, those gaunt and ungainly carrion eaters who towered over the other hunters, despite the feeding of hundreds of touracos and barbets, of chats and starlings, the caterpillar army had not been measurably diminished. Surfeited for the moment, the baboon had sat in a tree and seen the arrival of swallows on the fourth day of the feast; a thick swarm of migrants heading for feeding places in the southern continent where they expected seasonal rain.

Behind this first wave of swallows were hundreds of thousands of others, combined in one of their occasional mass migrations involving nearly all the birds who had come down from the northern hemisphere that year. They had been headed for a place precisely remembered from other seasons, a half-dried swamp surrounded by expansive grasslands where they always met rains which stimulated insect populations. The supply was big enough there to satisfy their mass appetite for nearly a month before they had to split up and move further south.

The caterpillar feast had stopped the swallow migration. Normally, the birds had no way of reaching the caterpillars since the insects were usually well concealed in thick grasses or among the cloaking foliage of trees. These caterpillars, however, were at the crucial second stage of their population explosion, their numbers straining the capacity of the water-hole territory to support them. They ate so ravenously and with such abandon that the swallows were able to swoop down as if they were drinking water and snip the caterpillars from their feeding stations. In this way, twenty or thirty million caterpillars died in the first hour of the swallow onslaught, the destruction only slowing when the food-stupefied birds were forced to seek rest in trees around the baboon. But digestion was swift, and flights of swallows left resting places continuously throughout the hours of daylight while, almost as continuously, new recruits joined the feast.

By the end of the sixth day when the caterpillars were approaching their moment of mass pupation, other hunters had joined the swallows. They announced themselves through a small advance group of migrant pratincoles. These birds were migrating south along roughly the same route as the swallows and for much the same reasons. The graceful, swallowlike birds resembled some seabirds, the terns, in appearance. Unlike the swallows, they did not move in mass congregations but were spread out across a hundred miles or more, each flight group hidden from the others and apparently independent in the migration.

The moving pratincoles were not isolated, however, but had devised a method of linkage which made them sensitive to food opportunities along the flight route. The first of the pratincoles arrived soon after midday, and they gorged, rested, and gorged again. By late afternoon, they were surfeited, and in gradually thickening numbers they took to the air, gripped in a gently rising thermal above the water hole. The movement quickened and became a funnel of rising birds. In their circling, they flashed the white undersides of wings to the setting sun, then showed the black topsides so that the sky was turned into a kaleidoscope of black-and-white spots whirling like a giant, ascending beacon. And a beacon it was.

The pratincoles were visible for miles in every direction, and their vaulting flights became a signal for distant pratincoles, who came pouring to the caterpillar feast. They were still arriving when darkness rose from the grass, and the soaring birds fluttered to earth. For the baboon and his troop, it had been an uncomfortable night. They were surrounded by thousands of pratincoles and swallows squabbling and squeaking and erupting into sudden alarms as they, in turn, attracted owls and night-hunting cats.

For the following two days, the signaling of the pratincoles had drawn in thousands more birds. At the height of the caterpillar hunt, the baboons had been excluded from the feeding area because the press of creatures had created a confusion that was too much for them. Seething congregations of birds dominated the water hole. Streams of birds soared into the air in perpetual signaling to other pratincoles. Floods of newcomers arrived every hour. The caterpillars humped along desperately, gobbling away as they were themselves gobbled up. The mathematical equation that had made them overwhelming had been turned against them. On the morning of the sixth day, just as the surviving caterpillars were preparing to pupate, the swallows had streamed away to the south in a dense cloud of birds. The pratincoles moved in and cleaned up the last of the survivors. The water hole was ravaged, its vegetation dead and spattered with the white specks of bird guano. Stupefied storks stood around like sentinels, while kestrels and falcons dumped their bloated bodies in any tree where they could roost. The territory of the caterpillar feast had become eerily silent, the massacre concluded, and no moths had flown onward from that point of disaster.

Now, however, sitting hunched and alone in the vastness of a serene green landscape, the baboon munched moths contentedly. The high danger point of early morning had passed. He knew the hunting dogs would not be active in the hot midday. The lions would be asleep and so would the cheetahs. The leopard certainly would not be moving. He was as safe as he ever would be in this alien world. He had come on the moth congregation at its fringes and was eating his way into it.

Other colonies of safari ants were also searching for food. The safaris, unable to cover territory as easily as the baboon, had developed an almost perfect method for finding food. Each colony used food scouts who hunted constantly. Working alone, and therefore inconspicuously, they in no way suggested the throng they represented; they were secret agents making discreet inquiries. One of them climbed over the baboon's foot and passed on. Another investigated his rump and skirted it. Food scouts from one colony could cover acres of territory in a day, thus preserving the colony intact, leaving it ready for its great and ravenous expeditions to exploit concentrations of dependable food.

The baboon did not see or feel the ant who had climbed over his foot and was unaware of the scout's excitement when he found a moth—plucked up the next second in the fingers of the baboon. The ant's excitement grew when he found another moth, then another and another. Perhaps he scented the mass of moths, perhaps he was equipped with race knowledge that one of these moths meant many others.

155

Whatever the mechanism, the scout turned and fled through the grasses.

It did not matter how far the scouts strayed from the central colony because each creature laid down an individual scent which enabled him to return along his own tracks at any time. They could become lost only if their trails were obliterated, and this happened occasionally when larger animals made their kills, or when grazing antelopes and horses churned the ground betweeen them and their headquarters. Thousands of unsuccessful forays were made each day by the scouts, but such were the resources of the ants that these efforts were of no account. While the baboon ate on, oblivious, the successful scout stumbled and scrambled at frantic speed back along his scent trail, carrying the news of food. Within minutes of his arrival at headquarters, a column formed and prepared itself to follow the scent of the successful scout. It would follow his scent alone, even though it might cross the scent paths of a score of other scouts. As the baboon munched on, the great, coherent unit of searchers, hunters, transporters, guards, and workers that formed the safari society swung into movement toward him.

The baboon was closer to the ants in his own social scheme of things than almost any other animal in view at that moment. He, too, possessed the group idea of hunting, although he might not share his food with any other baboon. He, too, traveled in coherent units, classes of creatures encompassing the core of experienced leaders, the flanking guards and vanguards, the groups of juveniles, and the protected mothers and infants. But the vast gulf of brain and muscle that lay between ant and baboon was made incomprehensible when the two were directly compared. The safaris had by far the more complex system of life, by far the more organized method of gathering food.

The baboon was higher in the scale of evolution than the inscrutable ants; yet he was reduced by their efficient organization, and this only deepened the mystery. The safaris cut a path toward him, and he shifted position slightly, dropping his rump on the edge of a termite mound. He felt its soft warmth collapse under his weight.

The termites, almost as efficient as the ants, had been made aware of the baboon's presence by slight tremors communicated through the ground as he moved. When his jaws chomped, the action shook his spinal column so that his buttocks vibrated and sent signals into the earth. These were familiar warnings to the termites, but when the city walls crumbled, the reaction inside was extreme because there was no way to measure the scope of the damage or the nature of the danger. It was assumed that a major breach had occurred and that maximum effort would be needed to repair it. Perhaps antennae touched or heads tapped together; whatever the method, the information that a break had occurred was somehow transmitted to the deepest reaches of the metropolis almost instantly.

The termites lived in safari ant country, and the ants were their greatest enemies, so their response to any breach was immediately aggressive. Warrior termites rushed upward toward the baboon's buttocks. They belonged to a soldier caste and were precisely fitted to perform this one function. They had become such specialists that they had lost the capacity to take care of themselves. They were helpless dependents of the workers who fed and cared for them when they were not needed to fight. They were sterile and thus doubly dependent because they had no control over their numbers. These were determined by some mysterious intelligence that understood the colony must not be dominated by a military caste. Thus, when their warrior battalions reached full strength, the termites' surplus soldiers were killed or allowed to die.

Now, a thousand soldiers poured upward into the breach created by the baboon's buttocks. Like all the termites in the mound now, they were blind, but their other senses were acute. Their great jaws were designed to sink into an enemy and remain hooked there until the termite's head was pulled off or the enemy was dead. In some ways, therefore, this was a suicide squad going to the defense of the city. The swollen head of each warrior contained a gland that manufactured the termite's offensive weapon, a sticky substance that penetrated, clung, burned, and was strong enough when it was fired in a fine jet stream from near the termite's jaws to curl an ant into immediate agony and probably kill it. No invading army of ants hoping to breach the city walls had any chance of success as long as these formidable soldiers could get to the breach quickly enough.

When the warriors reached the open air, they found no enemy. The baboon had moved to reach for another moth, dragging his rump a few feet away. No sound of ants touched the defensive soldiers. Almost immediately, the first of the builder termites arrived. They came to the construction site magnificently equipped for their task. No particle of food was ever wasted in the termite city. Any worker could feed any other worker or colleague at any time by disgorging partially digested food from either end of his body into the mouths of the hungry. Because the process of digestion was never totally completed, it mattered little whether the diner was eating disgorged food or disgorged excrement. Both contained the nutrition he needed.

So, nothing was wasted inside the mound. Unhatched eggs were eaten and processed. The dead were eaten and processed. The involuntarily expelled excrement was eaten and processed. This processing continued until finally the worker's body contained nothing but a thick darkish paste. And this paste was the crucial material, the mortar of the black-mound termites.

The first of the workers to arrive immediately rearranged the debris of the breach into something approaching its former order, using the paste in their bodies as mortar. Meanwhile, other workers came from deeper in the mound where the soil was still damp. They carried loads of building materials which they piled at the breach in a long, continuous wall, all of it held together by the processed adhesive. While the repairs were being made, the soldiers stood guard duty, pincer shields open and swollen heads pointed blindly at the empty sky.

The volume of material pouring up from the depths increased; chunks of uneaten wood, gobbets of earth. The workers held these objects between their pincers and traveled with such speed that they shouldered each other aside in their eagerness to get to the work site. The reconstruction was so rapid that the baboon had not eaten a dozen moths before the breach was repaired. It was so mercilessly quick, indeed, that scores of builders working outside the breach, and soldiers still on guard when the last of it was patched, were left facing the blank, hostile castle; no openings anywhere. They ran about frantically once they realized their plight. Perhaps they heard, or smelled, the approaching safaris.

The afternoon sped by in peace, and its demise turned the baboon back toward the island. It was difficult to know when the hunters would begin their evening work. They were usually predictable, but sometimes a hungry cheetah hunted in the heat; sometimes a wandering lion might surprise the unwary. The baboon turned away from his feeding place, saw the safaris this time and stepped over the black column. He paused for a moment, debating whether to open the termite mound, then decided against it since his appetite had been so well satisfied with moths. He scampered toward the island.

The termite city went on with its work, while mites took advantage of its inhabitants. The mites were agile little creatures who could jump from one termite worker to another as they passed each other in narrow corridors, and the mites teemed throughout the city. Some crawled over the queen and her consort and sneaked down into the mouth region to steal food as it was placed there by the busy workers. Others scavenged ahead of the termite cleanup gangs who rooted through the city for any available detritus that might be turned into food. In their resting stage, the mites clung in masses to the workers and so were carried to every part of the city, but once they became active again, they dropped off their hosts and scurried about looking for scavenger material.

The mites were not the only strangers in the termite mound. Many kinds of tiny protozoans, creatures close to the creation of life itself, waited safe inside the bodies of the termites. Some day, if they were lucky, a hunter would break into the mound and then the termites would be helpless. They would become victims of their many enemies, particularly the bustards, the storks, and the other ground-hunting birds. Then the protozoans would be carried away in the intestines of the hunters, free to go ahead to the next stage of their development. Thus, what appeared to be that ultimate refuge on the plains, the termite city, was in reality a self-generating producer of life, spreading its influence far beyond the visible scope of its humble mound, yet never once betraying the importance of its function in that place on the plains.

IX

The African moonlight, made unique by the altitude of the land beneath it, was of a peculiar clarity, with a capacity to penetrate and illuminate, to spread impartially over every part of the grassland and woodland, over river and lake, mountain and valley, and the night world roused itself in response to the light. The moonlight stole among every blade of new grass, lightly touched every leaf in every angular acacia, and revealed a world where all was visible, yet only half-seen. This was the perfect night world of the hunt, where the cool, the calculating, the concentrated combined their energies to make spectacular kills. Moonlight shimmered the silent surfaces of streams caught in backwaters. It glowed on the flaring snouts of hippopotamuses waiting for their evening chance to leave the water to graze on dry land. It stole into the yellow eyes of the leopard, invisibly mottled on his low, strategic branch. It shaped the hunched form of the baboon, the curve of his back revealing the quiet agony of his effort to sleep in a place so well filled with danger.

Moonlight helped the evening hunters so much that some animals would not attempt kills without its assistance. But so subtle and quiet was its presence that it offered little help to the potential victims. It extended their vision a few score of feet, but when they were in panic-stricken flight, their hooves stomped down on earth too ill lit to reveal its manifold burrows and hollows, dens and refuges, and the sounds of snapping antelope legs were lost in the thunder of general flight.

At the edge of the woodlands, more than thirty miles from where the baboon sat uneasily asleep on his island, the moon sent a shaft of light down into a hole flanking an untidy mound of hastily thrown-up earth. It shone into the pallid disc of the old hyena's eye, luminescent as it caught the moon's rays, and moving slightly as it responded to the message of the light. The hyena's blind eye saw nothing of the moon, but her other eye, still hidden in shadow, blinked at the light. She groaned with the knowledge of the great effort that lay ahead of her. She was not by nature energetic. She exerted herself only as much as was necessary to survive. But now, survival was painful, difficult, and altogether unpredictable. She groaned again and expelled a whistling gasp of air.

She knew she must move. She was hungry. Now that the antelopes had mostly passed through her territory on their way to the short-grass plains, she was left with cubs in country that could scarcely support herself and her family. At the beginning of the rains, her territory had gradually elongated into a fluid line, conforming to the movement of the animals themselves, and she had gone slowly south, probing into new territory, meeting many strange hyenas. Sometimes, there had been no true territory, and she had joined with a score of other hyenas of various tribes to hunt briefly together. It had been an exciting time, a time of movement and thrilling chase, of fight and concentration. But the antelopes and horses were now well spread, and harder to find.

She heaved herself up and burst into the glow of moonlight with a clumsy sprawl of limbs. No part of her was beautiful. Her bowed, bulging belly hung down between her powerful legs. Her sides, spotted with the dark blotch marks of her kind, were scarred and torn by the hundreds of fights she had won. Her once-rounded ears were ripped and torn to tatters, and her right eye stared sightlessly into space, its scarred white disc cast upward in an oddly ferocious aspect. When she ran, her short, bushing tail curved tautly over her spine, her belly swung heavily from side to side, her clumsy legs galloped like a separate mechanism under her, and her teeth were bared over her lips. She looked uncompromisingly ugly.

Although her appearance was repulsive, each part of her body was adapted to exploit all the possibilities of the hunt. The ugly body and loose, swinging skin concealed thick, tough muscles which could drive her without slackening speed for more than ten miles in pursuit of an antelope. These muscles could send her more than fifty miles from her den in a single night in search of meat to charge her udders with milk for her youngsters.

Her eyesight was poor. Blind in one eye, her remaining eye was scarred and thickened with old injuries. But compensating for this was her acute hearing, accurate enough to hear the *shape* of activity in the darkness as she raced forward to make a kill. It was her hearing that instantly caught the sound of vultures coming down in the wind while she was dozing, the sound leading her at a fast clip to investigate the kill, perhaps a mile away.

But it was her stomach, above everything, that made her the most efficient killer and scavenger in this country of big animals and massive death. It accepted anything, everything, processing fur, hide, razor-sharp chunks of bone, meat ripped from the living animal, or meat so putrid that even vultures would not touch it. It was a stomach that could digest, or attempt to digest, ancient leather so hard it crackled like broken wood as it snapped in her teeth. It accepted chunks of wood and mouthsful of earth and dried grass. It was a two-directional stomach which could process gargantuan meals at fantastic speed and, conversely, throw up anything that did not agree with it. It collected hair from the hides of the slain and wrapped it into compact balls to be sent upward while the meat was being processed downward.

Her jaws were among the most ferocious hunting implements on plains and woodlands. These were jaws so powerful that once, baiting a lion, she had made a fast nip at his back leg and the big cat had limped for ten days. These were jaws that could smash the thick bone of a buffalo leg in half. They were jaws that could crack skulls, jaws with such gripping power that if she got a tooth driven into the hide of a zebra, she could half skin him with a wrench of her head.

Now, with the moon rising quickly and an intermittent wind blowing softly across her muzzle, she stood and waited for some signal to begin the night's work. Her cubs ruled her movements, but she remained a gregarious animal, partially dependent on the whims of other hyenas collected loosely under her dominion. Because of her age—she was nearly ten years old—she radiated a dominating air which combined her experience, strength, and determination. In the far distance, dim shapes moved steadily toward her. Two inferior males had anticipated her wakening and her desire to hunt and wanted to accompany her wherever she might decide to wander. Behind her, and beyond a line of acacias snaking along a chuckling watercourse, the single whoop of another hyena sounded, then the voice of one of her cubs weaned the previous year. Three more shapes appeared along a ridge and started down toward her. The evening gatherings of hyenas were so loose, so informal, that no strict rules applied. The female was a leader, but if she were not present, the animals dispersed to hunt alone or in groups of two and three. But the members of the tribe listened for each other's voices, recognized them, and were always quick to dash to the scene of a kill or a chase in this area.

The hyenas were creatures of territory, bound to it like lions but apparently able to move over greater areas. Because the female was the largest and most powerful within her territory, she dominated the other hyenas and was the head of a loosely organized tribe of animals joined spasmodically in their hunting and breeding but permanently united in defense of the territory against alien hyenas.

Ignoring the giggles and yips of young hyenas playing near her, she headed along a path flanking the creek bed. The other animals followed her in a straggling line. They knew she was a supremely good night hunter but they were not aware that night hunting was becoming an increasingly desperate time for her. Each attack seemed a little harder than the last. She was stumbling, even falling, in high-speed chases. Her head was still ringing from the kick of a zebra two nights before. Worse, she awoke now with joints and muscles stiff and her doubts at a peak about her ability to kill again.

The other hyenas strung out behind her, scuffling and thumping, and completely ignored the night noises around them—the scream of a disturbed baboon, the rustle of a tiny antelope running through nearby undergrowth, the mournful whooing of an owl. Silver light shone in the female hyena's good eye when she turned to survey the animals at her back. Two years before, leading a zebra attack on the great plains, she had been kicked insensible. For eight days she had crawled in the shelter of thornbushes, dreading the sound of lion, leopard, and wild dog. She had been left deeply uncertain about her power over the hyenas. She had become dominant not only because she was big and powerful—all female hyenas were bigger than any male—but because she had special skills in the hunt. Her terrible bites at tender zebra underparts often killed, even though her one hundred pounds of weight were dwarfed by the thousand pounds of her victim. The other hyenas acknowledged her supremacy and followed her.

She knew this territory intimately. She had ranged as far north as the dry foothills where hunting was poor, and as far south as the end of the plains themselves where other foothills began climbing into the mountains. She knew the water holes, the periodic-flowing rivers, the rock islands dotted across the flat grasslands. She could measure the distance between her den and the place where the grazing animals were feeding. Because she had wandered widely during her long life, taking territories in both woodlands and grasslands, she was able to understand something of the movements of the animals and their relationships to each other and to her.

She knew where each of the lion prides around her territory was placed, and she expected them to be fixed there from season to season. She was less certain of the position of the nomad lions, those young and old animals who might appear at any time anywhere, those perennial wanderers in search of food or companionship, of pride ownership or sexual gratification. The enmity she and her kind felt for lions was unique. The hyenas baited the lions, provoked them, bit them, badgered them. Their giggling, squealing, whooping cries sometimes drove hungry lions berserk as they tried to eat a fresh kill and were not allowed to. The lions responded by killing any hyena they could catch, sometimes dashing into feeding packs in the darkness and crippling half a dozen animals before the hyenas could flee. A lion could scrunch a hyena skull flat in one bite, then contemptuously toss aside the carcass. The hyenas, in turn, watched the lions for signs of sickness, injury, or other distress. The female had waited two days with other tribe members around an old lion groaning away his last few days of weakness. Then one night, in a quick, concerted attack, they had killed him. No hyena would choose lion meat in preference to the meat of antelope, but in this instance, no part of the big cat was preserved. In the morning, jackals sniffed the churned-up earth, the long blunt claws scattered at random, the pieces of shaggy mane strewn over half an acre of the plains.

But now, an immediate and critical matter was her present hunger. It was bearable enough when she was alone because she was well fitted to stand both shortages and great abundance of food. But her cubs were still on milk, voraciously so, and she could only produce milk if she ate a continuous diet of red meat. She must decide where to hunt. Hunting in her den territory was an uncertain affair in this season, since the grazing animals were widely scattered and could not easily be found. The impalas, collected in their breeding herds, were difficult enough prey anyway, with their confusing giant leaps and their great bursts of speed. The buffaloes, located in one large herd north of her territory, had massed together on rich green meadows in a hollow between two low hills. Any attack on them would invite the response of the entire herd, and few hyenas were willing to risk that. Dispersed throughout the trees were some bushbucks, wary and difficult to flush from cover, some topis who had not yet moved down toward the plains, and odd groups of elands. The elands were favorite eating, but they, too, were difficult to hunt. Their great size—they were by far the largest of all the antelopes—made it difficult to kill them. The easiest meat to hunt was far south, on the plains.

Thrusting south with more purpose, the old female increased the pace, and the extra effort silenced her followers. Some of the younger, less experienced animals, hoping for quicker kills, dropped away and were lost in the gloom. Behind the old female ran other mature females, followed by the most dominant of the males, then by the younger animals. At the rear—and she did not dwell on this fact—were the very old, a male and two females, struggling to keep up with the rest of the tribe. When she stopped at a small watercourse to scan the still water for a sign of crocodile and to sniff for the scent of hippopotamus, she could hear the labored breathing of the oldsters behind her. After a moment's hesitation, she plunged into the shallow water.

The road south rolled on in the light of the moon. Keeping up a steady pace, the tribe passed through the region of thinning trees, crossed the last intersecting river, skirted the beginnings of rock islands, and so entered the territory of the plains proper. Now, the female knew, she was within striking range of the grazing animals, but she had no immediate knowledge of where they were. The hyenas might traverse almost the full length of the plains, an all-night run, and not encounter a single animal, or they might not be able to move more than a thousand paces in any direction without colliding with large herds.

This night, the thrusting hyenas almost immediately heard zebras returning north to clean up a mixture of short and long grasses which had sprouted again after being grazed down in the first flush of migration. The female did not change the tempo of her stolid running, but mysteriously, the running transformed itself into a determined attack. Some hyenas, aroused by the changed purpose of the lead runner, yelped and chittered their excitement. This drew other hyenas out of the gloom so that the pack which had begun as fewer than a score of animals now numbered nearly forty. The heavy slapping of their paws sounded against the breathing presence of the zebras, still unseen in the gloom ahead.

At first, the zebras, fat bodies loose and ears upright,

cantered easily away from the hyenas, but the hyenas gradually overtook them as the energy of the zebras waned, and the chase abruptly became tense. As the hyenas closed in, stallion after stallion dropped to the back of the herd to prepare for defense. The female hyena, well in the lead, ventured too close to the last stallion. He kicked her at full gallop in the chest and face and she suddenly shot upward, turning over in midair. She landed with a hollow crunch among the hyenas following her, bringing two of them down with her. She felt the wrenching strain on her back and hindquarters as she fell, but the stallion's blow had not been full force and feeling no real pain, she rolled to her feet and was running again.

Now, the hyenas were strung out across the full width of the fleeing zebras, dodging kick after kick as the stallions sought to keep up the pace of the flight while trying to blunt the hyena attack.

The sounds of the hunt dominated the night. Two hundred hooves hammered across the plains, and the air was filled with squealing zebra cries, punctuated by the hoots and giggles of the hyenas. Ears back and teeth bared, the stallions skidded to stops, biting and lunging and kicking at the hyenas. The attackers howled when they lost the tips of ears, were kicked in their sagging stomachs, or were trampled by panicky zebras trying to rejoin their families. A young male hyena was kicked and his body catapulted vertically, his moans dying away as others charged ahead into the zebras and around their milling masses.

News of the attack fled across the plains faster than the footsteps of any zebra. It told the leopard the position of the hyenas and warned him to avoid any hunting attempt in open country. Lions listened intently, each nuance of sound registering in their heads so accurately that the squeals of the kicked hyena sent one of the young lions starting to his feet as he sensed trouble. But nothing the lions heard stimulated them to action. There was no ghastly scream of a stricken zebra, a cry that would have offered them red meat.

Despite persistent running, the old female could feel that the attack was a failure, even as she flanked the uncompromising resistance of the zebras. She missed a lethal thrust under the belly of one mare, and gasping, she turned away. Her retreat blunted the enthusiasm of the others. Limping hyenas retreated with her and collapsed, panting. Gradually, the sounds of the fight died away. Slowly, the zebras relaxed and formed up in the moonlight. They walked easily together once more, each stallion in his place, each animal knowing that hyenas would not attack again that night. They walked past the hyenas and disappeared to the west.

In the long experience of the female hyena, only one hunt in three succeeded, and this only in good times. When drought or disease stalked the hunting grounds, a dozen hunts might bring no food. Now, she felt the pains of the kick she had received. Once, she would have rested hardly at all before being away again, but this night she lay exhausted. A silent hour passed. Then a distant cry, pounding hoofbeats, and a howl told her that one of the other hyenas had wandered to attack a herd of wildebeests. Hyenas started to their feet all around her, and in seconds they were all pouring through the drenching moonlight toward the sounds of the chase.

The running reminded the old hyena of how far she had come from the den. A year before, she would not have felt these pains or such ferocious hunger. A year ago, she would have been totally directed toward the kill. But now, the pains of old limbs interposed themselves between her and her goal. Worst of all, she was running toward a hunt started by another hyena, and as a result, she might not be able to maintain her position at the head of the running pack. When the moon slipped into the refuge of a cloud, experience cautioned her to slow. No creature could run safely at top speed over earth littered with termite mounds and the wreckage of their abandoned remains, earth that was pitted with the holes of bat-eared foxes, jackals, honey ratels, other hyenas, and warthogs. One misstep into a mongoose burrow could break a hyena's leg just as surely as the leg of a fleeing antelope.

The old female remembered one night when she had recklessly led her tribe in pursuit of a wildebeest cut from his herd. She had been pressing the attack so hard that she had burst out of a thicket to find herself falling to a jolting landing on top of the wildebeest, himself rising from a pool of dark water. Hyenas had poured into the water after her. But she had remembered this was a crocodile place where sometimes long, sinister drag marks touched its mud and sand shores and where a blazing green orb sometimes watched her from the far side of the pool when she risked drinking there. Knowing that the crocodile must still be in the pool, she had driven herself desperately over the wriggling bodies of her comrades, kicking them deeper into the water as she boosted herself to land. Standing there, panting in the gloom, she had heard the others trying to finish off the wildebeest, the sounds of squabbling almost irresistible to her. But then had come a great splash which had sent all the hyenas churning for safety. The strike of the crocodile was

173

made so swiftly that no animal could know whether it had killed or not. The wildebeest's head had reared up, mouth agape, and then he had disappeared underwater as if drawn down by an invisible and irresistible force.

Inhibited by memory and the weight of her experience, she fell back now, her heedless companions pounding steadily past her until she was at the tail of the pack. She had always possessed that separation of brain and body which made her cunning, and aware, and capable of learning. This was a tradition of her race, as sharply developed when she had been a cub as it was today in her old age. During her own cub life, she had starved for days while her mother hunted, but her hearing had been so acute and selective that she had been able to tell the difference between the footsteps of her mother and those of male hyenas. This acuity had once saved her life when a hungry male had come on the den. He had turned up her brother and sister and killed them, and only missed seizing her because she had dug so early and so quickly away from the danger she had suspected from the footsteps.

Even now, with the breath gushing in her throat, she heard far beyond the racket of the advancing tribe. The clumsy pounding of hyena feet, the jolting of her head, and the panting of other breaths did not diminish the continual flow of information she was receiving as she ran. On her right, she heard the steady clomp-clomp-clomp of a nomadic lion who ignored the hyenas and was heading for the woodlands. The old female marked his position and linked him to an encounter with a young nomad, deeper into the plains, a few nights before. Nomads disturbed her since they were so unpredictable and did not advertise their presence with the communal roarings of pride animals. The clomping footsteps died away, and the old hyena dropped further back. Now, she could hear the impatient stamping of wildebeest feet ahead as the animals, parts of a now-fragmented herd, divined the approach of the main hyena pack. The original attackers, lacking experience and skill, had driven themselves wildly into the center of the herd, splitting it into three sections which broke away and wheeled in broad circles eventually to come back together again.

The attack made no sense to the old hyena, and she had no heart for it. This was not an incisive collision of hunter and victim, the appearance of the hyenas abrupt and perhaps unexpected. There was no quick panic of the hunted and their furious galloping flight, with the hyenas loping easily and spreading out as they became more certain of their victim while the herd determination weakened and the animals failed either to escape or turn the mass of their numbers to useful defense. Instead, this night, the hyenas themselves were scattered, and even as she ran, the old female could see and hear the pack ahead of her spreading out too far.

Thus, she was well behind the main body of hunters when the hyenas collided with the wildebeests. She heard hooves crashing the giggles and whoops of excited hyenas as she slowed, then stopped and lay down. The attack came to her through a series of audio messages transmitted through the moonlight as clearly as if she had been seeing the episode in the light of day. The original two attackers, exhausted by their failure to kill, had circled on the heels of the fragmented wildebeests and come in at the flanks of the new attackers. Their exuberance was blunted, but their appearance stimulated the foolishness of the main body of animals. Instead of hearing a rising thunder of fleeing feet, the old female heard a slackening of movement, the feet of the hyenas sounding louder than the hooves of the wildebeests, and she divined that the wildebeests had stopped running. They had gathered into a large group, or several large groups, as was their custom when they sensed that attacking animals were not truly determined. The bulls were to the front, with their heads down, and the cows behind them, while they waited en masse to see whether the hyenas would charge. The running hyenas chose, instead, to circle the grouped wildebeests, a maneuver which had no ending because the longer the circling continued, the more certain the wildebeests became that they would not be attacked. The old hyena waited, and although she made no sound, the others must have marked her position, because in time, they dropped away from the attack one by one and joined her, slumping down around her in a large circle. Without moving, she was once again at the center of the group.

The night was half-gone, the moon now almost directly overhead, and the old female could not feel any lessening of the apprehension that had started her evening. She was now so ravenous that she had difficulty in remaining still. She wanted to be up and running, scenting the soft night wind, ears pricked for sounds of red meat to be gulped into her now-rumbling belly. But matching this intuitive urge was the cautionary warning of her old bones, the discomfort she felt with so many other animals around her. Although she was dominant among them, she felt acutely uncomfortable when too many wanted to be with her. Her experience of mass hyena hunts was negative. There was rarely enough meat to feed all the mouths, and the ferocious competition at the carcass was physically dangerous. Sometimes, hyenas themselves were eaten, their deaths unmarked in the heaving throng of bodies smothering the first victim. She, herself, had been nearly crippled several times when she had been bitten at such feasts.

Her discomfort grew, and so she got up, yawned, stretched, and wandered back and forth, sniffing the ground. Her actions were apparently aimless, but she moved steadily beyond the grouping of hyenas. Some of the others watched her languidly, but not sensing any resolution in her behavior, they did not attempt to follow her. When she was out of sight, she began a long, loping run which would take her due south, deeper into the plains. She knew this territory as well as the country around her den. On her last trip here, it had been a country of gazelles, and now she found the gazelles were still here. In their customary fashion, they were following the zebras and wildebeests, who had reduced the grasses to a height that made it convenient for them to eat.

She was concerned that she might be tempted to make a run for a gazelle when her real appetite now was for a larger animal. Her secondary concern was a kind of lurking uneasiness at being among so many gazelles. In their mysterious way, they knew she was not going to hunt them, and so they felt no fear in approaching her. Many of them had the temerity to walk close beside her, within a couple of bounds, their stiff-stepped movements and flipping tails near enough to tempt her to a sudden charge. But she controlled the impulse.

She could not know that she was being escorted through gazelle country as she passed through the territories of individual bucks. Sometimes, she was between territories, and she trotted for several minutes without meeting any gazelles. But then, always, another buck would approach her at the limit of his territory, warily keeping his distance as he divined her intentions. He became bolder when he discovered that she was just traveling, and flanked her through his territory. Another buck appeared immediately as the first buck stopped.

Buck after buck took over his escort duty as she clumsy-footed along toward her distant destination. Not for a moment was she out of the surveillance of the territory-holding bucks, and at no time did she display any threat toward them. Nevertheless, she kept constant watch for any sign of the sick, the crippled, the uncertain. One limping gazelle was often enough to send her wheeling away for an easy kill, but as long as she was heading for the hordes of animals in the south, a few mouthsful of gazelle meat were hardly worth the furious and exhausting chase they always demanded, so she stoically endured escort after escort while the moon swung overhead and the rumble of lions darkened the distance ahead.

During the abortive attack on the wildebeests and her trek south, she had been listening to the voices of the lions and reading messages into them. Despite her loathing of lions, she was drawn to them in some kind of deadly contradiction of impulse that for many hyenas ended in unnecessary death. She could not consider lions without old memories making her tail bush out and curve over her back. Lions had once killed a big eland almost on top of her cub-filled den, and had taken five leisurely days to eat it. She had been forced to circle the bloated group, listening to her cubs crying hunger, until one by one their voices became silent. Lions did not pose any special threat to her, apart from the fact that lion and hyena were always in competition with each other. The hunting methods of the two animals precluded the necessity for direct confrontation, but as if by some tribal memory, she seemed bound to provoke such confrontations and to place herself in positions of unreasonable danger whenever she came into contact with lions.

Her rage at them, and her terror, the ambivalence of her reactions, did not stop her from following them and waiting for fifty or sixty hours, if necessary, for scraps of meat from their kills. Her emotions did not stop her from annoying them—less frequently now that she was old—in what seemed to be suicidal provocations. Inside her was a thread of insanity, present in all hyenas, which was intensified when she became involved with lions.

Ahead of her, two prides were working adjacent territories, and one of them, if she judged correctly, had meat in its territory while the other did not. Along the division line of the two territories, the lions rumbled and roared challenges at each other, and they transmitted their tension to all who could hear their voices. The female hyena turned slightly east of the sounds to flank the prides in whose territories, she knew, there was little opportunity for a lone hyena. And so, under the moving moon, she clumped steadily toward a looming line of trees along a watercourse. The solid, chunky shape of a plains island showed ahead. The rumble of lions sank into silence behind her.

In two hours, she had plunged deeply into the plains, into the seemingly trackless world of the migrant grazers, where her only landmarks were the rocky islands that appeared ghostlike in the moonlight, and the intersecting scent trails of zebra and wildebeest, cat and dog. Once, she stopped in puzzlement, the scent of leopard strong in her nostrils. It was accompanied by the disturbing smell of violence, of blood in the air. She was not to know that the wounded leopard had just minutes before essayed his first trip away from the island. She was not to know that he had pushed north toward the familiar cluster of islands dominated by the territorial leopard, and had found himself caught in a pincer of opposing lion prides. Injured and weak, he was pinned down in the open grasslands, his only protection a feeble tree that would scarcely support the weight of a baboon.

While she hesitated, a zebra's whistling alarm call brought a response of galloping feet. A brief silence followed. Another alarm call sounded. Furious galloping followed, then a violent crash, high-pitched screaming, and the thrashing of limbs.

These were the worst possible danger signals for the leopard since they indicated a confrontation between three groups of lions—the two ahead of him and the group that had just audibly made a kill behind him. He was not to know that the southern kill had been made by an old nomadic lion who had been tracking the zebras for most of the night. For the hyena, the sounds were an invitation. She rushed forward. The screaming stopped as she was running. In the darkness, the lion's teeth had gripped the zebra's nose, and the big horse was slowly suffocating. The old female heard other hyenas communicating in the gloom as they, too, moved toward the kill.

When she reached the stricken zebra, other hyenas flanked her. More were appearing out of the darkness, and within minutes the lion was encircled with his kill. The presence of the other hyenas at first dismayed the old female, since they were all strangers, members of a tribe she had never before met, but in a moment her old feelings of dominance rushed to the front of her awareness, and she felt the need to lead, to direct her fellow hyenas into decisive action.

Fur bristling, tail curved tensely, she began a slow, half-sideways advance on the lion, whooping and uttering low snarls. The lion ignored her. She bared her teeth, scarcely six feet from the big cat now. He brought up a menacing rumble from deep in his throat; she was too close. He got up. Instantly, the female darted forward, gripped a piece of zebra meat and ripped it free. The lion's great flat paw came whacking down—but she had wheeled away, gulping and choking over the chunk of meat. The moon disappeared, and hunters and hunted became dusky silhouettes. The crouched leopard, waiting for this darkness, made his sprint for the trees, still possessed by the erroneous idea that he was trapped in the middle of a large pride of lions. His silent disappearance registered on no other creature, and his shadowy flight into the canopy of a tree was not seen.

The success of the old female's bravado brought the other hyenas closer. The chorus of their cries became deafening. When the lion rushed at her, a dozen hyenas grabbed mouthsful of meat. The lion whirled to drive them off, but the female bit his leg to the bone. He roared, and caught one of the other hyenas a glancing blow that bared white ribs. He charged back to the carcass, now covered with giggling, whooping hyenas whose cries seemed to have a scornful ring to them.

Again, the sounds of the conflict radiated and had their multiform effect. Jackals paused and listened, then decided

181

that the killing was successful and that there was conflict over the meat. There was always a chance for food at these contested kills. They began running toward the distant roarings and whoopings. Elsewhere, other lion prides listened intently to the sounds as they sought to establish whether this was a pride kill, who had made it, whether it was contested, and what chance there might be for a quick, opportunistic intervention. Some young lions in country that bordered the territory of the kill became highly agitated as they felt the pull of the fighting urge to preserve this country for themselves.

But the older lions held still, understanding that with such a large number of hyenas present, the chances of easy piracy were remote. Hidden in dark treetops, storks stirred to the sounds of the conflict and marked its position for investigation in the morning, knowing that fierce fights over prey produced untidy and disorganized kills where many scraps of meat were left. Their long, probing beaks would miss none of them. The vultures, too, registered the fight, but they were less affected. Despite the nearness of morning, the air around them was still heavy with humidity, and it would be midmorning before they could hope to take to the air, far too late to share in any such kill. The baboon, now thoroughly aroused, listened to the muffled sounds of action, and he trembled.

The messages from the kill passed back and forth, and ultimately, the information would arouse the creeping, crawling creatures of the subterranean world, if it had not already done so. Ants would form into columns to reach the scene. Beetles would gather by the hundreds, so that eventually, no scrap of fat or hide or even hair would remain to mark the death of a zebra at that place.

But the zebra was not yet dead. His screams had been silenced by that awful grip over his nose, but it had not quite killed him. Despite his injuries, he was gulping fresh air into his lungs, gulping life through his mouth while his life's blood flowed out of a dozen wounds elsewhere on his body. The lion was now too engaged to do anything more than fend off the bold hyenas. He swiped the empty air again and again and was rewarded with swift and savage bites, so that his hind legs streamed with blood. Behind him, the zebra reared up, ripped and torn, but making his final effort to escape from the impossible, a ring of death from which there could be no escape.

The lion turned back for a second and saw his prey apparently escaping. He leaped on him, his powerful back legs splayed to get the grip necessary to knock the horse down finally. But instantly, both his back legs were smothered, as if by leeches. The hyenas ripped and tore at him and this was too much for the lion, who faced the very real possibility of being crippled or killed by these relentless enemies. With a final roar, he jumped clean over the zebra, knocking down several hyenas, and bounded away into the darkness.

Now, every hyena became an enemy of every other one. The old female, struggling to reach the victim she had been bold enough to make available to the mass around her, was buoyed up into the air, her feet kicking, by the press of animals on either side of her. She struggled to reach the ground, but this was impossible, so she tried to get above the other hyenas, and so found herself standing on their writhing backs. The zebra, still alive, looked at her from inches away, then brayed his last moments of life as his hindquarters were eaten, as evisceration was completed to the sounds of ghastly ripping and slavering noises from the center of the hyena pack. As his eyes glazed and his head went down, the old female was already eating his neck. The intensity of her feeding was increased by the knowledge that in fewer than fifteen minutes the zebra would be gone, the grass licked clean, all the large bones scrunched and swallowed. She must fill herself now or perhaps lose her cubs.

Shortly after dawn touched the eastern sky coral pink and the harsh cries of bustards rasped in the still air, the old female lay half-hidden in a pool of mud left behind by the advancing armies of zebra who had passed this place in previous days. She was too bloated with meat to travel at all. She had eaten well because, standing on top of the carcass, she had been able to eat her way down while simultaneously biting at the heedless competitors beneath her. She would digest this great meal, make the milk, and then begin the long run home.

Her solitary brown eye watched the sun leap into the golden sky, saw vultures making their giddy climbs in the still clearness of the morning. It was going to be a good morning for flying after all. Smaller vultures were circling toward the scene of the kill. The baboon watched the vultures apprehensively. They signified the presence of the big meat hunters, but even after the most careful night listening, he had little notion of who was out there or whether they would come to the island. Storks pumped overhead on their way to investigate what they had heard so loudly the previous night. The hyena's eye closed. The perils of the night dissolved into sleep, and in her dreams she became young.

X

Heat rose from the grasslands. Toppling clouds marked the sky above a narrow line of hills on the horizon, but neither thunderclap nor rush of rain diminished the force of the sun. The days repeated themselves, exact copies of each other, and the twin worlds of plain and woodland steamed and dried and pushed forth new life between the rains. Many of the larger animals sought shade, but the birds thrived in the intense heat. Sunbirds darted into the heavy air. They lived for the sun, snapping tiny insects from flower petals, fitfully sucking nectar, and pausing on twigs to reveal the exotic details of their iridescent reds and greens and yellows and blacks.

As though banished from the scene by an invisible arbiter, the sunbirds suddenly were gone, to be replaced by rollers, who climbed steeply into cooler blue skies, then fell in a series of spectacular rolls, tumbling end over end toward the earth. They, too, abruptly vanished, and the brilliant bee eaters appeared, groups of them showing bright color as they darted and hovered and snapped at insects disturbed by the feet and nibbling jaws of grazing antelope and horse, rhinoceros and elephant.

The rains had begun the transformation of many of the birds, galvanizing them into myriad activities, changing their colors and their moods. No transformation had been more startling than that of the widow birds. They had been dun-colored, sparrowlike birds with short tails, so ordinary looking that they were lost among the thousands of brilliantly colored creatures around them. When the first rains of the new year fell, the male widow birds had molted almost immediately. Pure black feathers had patched their bodies, then spread and covered, or banished, the undistinguished gray feathers. Their short tails had fallen off, and magnificent long plumes had grown instead. Their beaks had thickened and turned from an anonymous horn color to a vivid white which contrasted with the new black raiment behind. This change of plumage had occurred while the widow birds flew in gregarious hunting flocks which looked and acted the same. But the transformation had dissolved the convivial groups. The earliest full-plumaged males had dropped away from the flocks to begin the long and difficult business of staking out their territories for the breeding season that now lay just ahead. Each male had retreated to a chosen place, its significance known only to him, but usually marked by a half-overgrown ring of bare brown earth decorated with a single tussock in the middle. This tuft of grass was oval with a recess at each end, and the grass stood about eight inches high.

Male after male left the widow-bird flocks to take up new territories among the rain-long grasses. Each worked hard to rip up any grass growing in the circle that surrounded the middle tussock. The leaps became so steady and wide-spread that a score of dancing birds appeared almost any-where that hyena or lion, antelope or horse might look. Some birds had established two or three rings and leaped from one to the other, squabbling briefly with nearby males who infringed on territories now big enough to be difficult to defend. The owner of a three-ringed territory was forced to dance more than his poorer fellows for his display to be effective.

The dancing was not a simple matter of flying boldly to express desire, as it was with some of the other displaying birds on the plains and in the woodlands. Wings were not used at all, so the jumps were enormous for such small creatures. Two-foot leaps sent each bird upward with his neck arched back and his feathers fully expanded while the long-plumed tail curved two ways, one set of feathers point-ing toward the bird's head while another set dropped out-ward toward the ground. Wings quivered and tiny legs flailed in aerial running as each bird uttered a series of muted, metallic cries.

The dancing exhausted the birds, who were compelled to rest after every six to ten leaps. But even on the ground their need to display continued. They struggled with flam-boyant tail movements, expanded their neck feathers, walked jauntily within the ring, turned toward the oval tuft of grass in the middle and postured before it, bowing and picking up scraps of grass and rearranging them carefully in the recesses at each end of the tuft. It was almost an act of nest building as they pushed their chests into these places and wriggled their bodies to smooth down the grasses.

The dancing went on as more males joined the display, making their tentative, initial leaps while the first of the females began responding to the importunings of the original dancers. The arrival of the females sent the males into a frenzy. Their bodies shook. They hissed and uttered their metallic cries. They circled the tufts, and their concentration was so intense they seemed about to burst with their passion. The females ignored them at first, flying off and leaving behind deflated, crestfallen males with flattened plumage and limp plumes.

Buried somewhere deep in this tiny world of the dancing widow birds were invisible signals understood only by their own kind. Some of the dancers bobbed up and down frantically day after day and never attracted a female. Others received a succession of females and mated with them; yet there was no external difference to mark the great success of one bird against the failure of another.

Oddly, all the females were mature and fertilized within a few days, while some males were still maturing to the point of mating. The females left the displaying grounds and started the construction of scores of nests placed together in tufts of grass. Each one was woven tightly in a compact ball fastened to living grass and was equipped with an entrance on one side. The males were left alone, many of them unbred. Some still danced on in a forlorn attempt to attract females who would not now come, but they quickly lost their passion and all desire to display. They flew off to resume flock hunting and molt out of their mating plumage, disinterested now in the nests or dancing grounds that had so recently obsessed them.

The dancing widow birds passed on, but they were only a single entity in the flush of feathered lives visible everywhere in plain and woodland country. Hardly any patch of ground remained untraveled by guinea fowls and spur fowls, by scuttling quails and larks, by pipits and coursers, by kestrels and falcons, by day-hunting owls and ostriches, by secretary birds and crowned cranes, by saddle-billed and marabou storks. The ungainly foot of the marabou imprinted the earth, and a second later, the tiny cordon bleu finch inspected the disturbed soil for a sign of food.

The spectacular unfolding of the broad wings of the crowned cranes was an explosion of white set against green, then against blue, as they rose gracefully and easily into the perennial wind. Pipits stuffed their beaks with green caterpillars and mashed insects and ducked into grass clumps where nests brimmed with youngsters. The extraordinary secretary bird gripped a mass of grass for her nest with one powerful claw and somersaulted backward, the power of the jump wrenching the grass free. But sometimes the grasses were tough and the somersault turned into an ignominious crash as the bird landed flat on her back. Such accidents did nothing to diminish the bird's need to somersault, which was, in a bizarre way, indisputably related to the delicate dancing of the widow birds.

The mating season revealed animals at the height of their primeval passions. The grumbling lions personified these dark emotions in their uproarious and grossly repetitive matings, but the secretary birds conjured new personalities into being. They ceased to act like birds. They growled like hungry wolves, the noise coming loud and harsh from deep inside their throats. They were odd occupants of the grass-lands at any time, walking with stalking dignity and assur-ance, their long, storklike legs sprouting underneath black knickers, their gray bodies lean and lethal-looking, their long crest feathers protruding backward from their heads. Everything about them indicated an evolutionary midpoint between the harmless and the dangerous. Of all the bird-hunters on the plains, only the secretary birds confounded every convention of behavior. Their aberrations were some-times so marked the birds seemed to be slightly mad.

When the baboon ventured south of the island in pursuit of caterpillars infesting the grasses, he paused at the sight of a secretary bird strolling slowly through the new grass and peering intently downward for the same caterpillars that the baboon himself was seeking. In a movement so violent that the baboon fled in terror, the secretary bird went ber-serk. With wings and legs akimbo, the bird bolted from the jerking flight of a butterfly. Then he turned, contorting his body into astonishing angles, and came at the butterfly. He chased the insect for a hundred feet, zigzagging at ferocious speed, with the butterfly in full flight from its mad pursuer. But as quickly as the chase had begun, it ended. The ba-boon looked back and saw the bird strolling slowly for-ward. A second butterfly, identical to the first, flew close to his head, but it did not concern him at all.

The baboon watched the secretary bird because he re-membered its talent for uncovering unexpected sources of food, particularly when it was prowling the woodland coun-try where the baboon rightfully belonged. There, the watchful baboon had often seen the big bird kick the half-rotted remains of a fallen tree trunk, smashing it to pieces. The destruction frequently uncovered lizards which the bird gobbled down, or swarms of beetles and centipedes which the bird picked over carefully and swallowed one by one until the baboon drove it off to feast on the insects himself.

The baboon knew that the secretary bird also had a special capacity for finding quail, which was one of the baboon's favorite foods, though he rarely ever had the luck to eat one. The quails crept through the long grass and were able to rise quickly at the first sign of danger. They could be effectively hunted only by secretary birds, falcons, eagles, and kestrels. Surrounded by these winged predators, the quails preferred to skulk where their perfect camouflage melded them into the background. But once lifted in the beak of a secretary bird, they, too, might end in the baboon's stomach when his quick charge deprived the big bird of its food. So now, crouched and waiting, he watched the long-legged creature.

The secretary bird's eyesight was as keen as an eagle's, and his fleetness of foot was almost equal to that of a bolting gazelle, so that to surprise such a creature, the baboon had to be quick and knowledgeable. Now, he watched the bird stretch its neck up very high. Standing up himself, he could see mongooses playing between two termite mounds that had been converted into their burrows. The mounds were well separated, and this was significant, as both secretary bird and baboon understood immediately. The bird waited until one of the mongooses had completed about one third of a quick dash between the mounds, and then he charged forward.

The mongoose saw him and for a fateful second vacillated between running for the distant mound and turning back. The delay was critical. He decided to turn back, but his pursuer was on him with a flying leap and a kick that sent him spinning. Now, both hunter and hunted were equal. The screaming mongoose was as quick to dodge as the secretary bird was fast to jump. A second lunge missed the mongoose, but a third jump caught him just as he reached the hole of the burrow. This kick sent him flying away from the burrow. Now, the secretary bird stood between him and his refuge. He had to attempt a desperate run for the second mound. The bird jumped, then jumped again. The squeals of the mongoose ceased, and the slight, limp body lay still under the claws of the tall bird.

Only a few days before, this same mongoose had danced in front of a group of guinea fowls. The dancing had been closer, perhaps, to pure gymnastics, the lithe creature jumping upright, rolling, twirling, his body twisting in tight contortions. The effect was hypnotic, and the guinea fowls had stood totally still as the mongoose had come steadily closer. When he had been at the correct distance, his display had stopped, and he had streaked forward to clamp himself to the head of the largest bird. The spell had been broken in an uproar of cackling, but the victim was dead with mongoose teeth in his skull.

The guinea fowls walked through long grass and short, through shaded fever-tree country, across sun-scorched plains, through dense rain that tossed mud on their wings as it smashed down on bare soil. Yet nothing seemed to diminish their numbers. Their cackles of outrage as leopards drove them out of trees, their uproarious disapproval at the appearance of the serval cat, a great enemy, were merely counterpoints to the special kind of survival that persisted while they learned nothing.

The abundant guinea fowls were spread widely and en-

dured as a constant resource, always available to the determined and hungry hunter. They provided predictable victims along the edges of water holes and watercourses, at the
rims of lakes, in the tops of trees, and in long grass everywhere. They had come one evening to drink at a temporary
pool near the baboon's sleeping place on the island. They
all had their beaks down when the baboon had swung out of
his tree in the dappled shadows. One swift rush took him
into the shallows. He had seized a guinea fowl and returned
to the tree pursued by futile cackling.

Now, the hungry baboon was stabbed by the sharp memory of another favorite food which, instinct told him, was
presently ripening in his true territory. The memory revived such a familiar and convivial time that he whimpered
his frustration, hearing in imagination the chomping of baboon jaws and the cries of youngsters playing in high
branches. This was the time of the ripening of commiphora, plumlike fruits about half an inch long which
attracted hordes of creatures. Normally, at this season, he
would be among the commiphoras in the woodland country, along with other troops of baboons, some vervet monkeys, and innumerable birds. The bright, flashy colors of
many different starlings would glint in the sun, while spar-
rows and social weavers, petronias and waxbills, shrikes
and flycatchers moved quickly among the trees.

The attraction of the commiphoras was a contradiction in
the world of the birds because the fruit brought together
creatures of such different appetites. The omnivorous starlings were to be expected, but the carnivorous shrikes normally did not eat fruit, and the flycatchers were entirely
insectivorous, except for their love of this one fruit. With
swift, circular movements of their beaks, they deftly re-

moved the outer flesh of the fruit, which fell to earth where greedy young baboons pounced on it. Then the flycatchers ate the inner flesh away from the stone, leaving the stone attached to the tree. The starlings, though, shared nothing with the other creatures. They ate the outer covering of fruit flesh, then swallowed the stones whole. Later, most often just before they took an evening drink, they would spit out the stones, now cleansed of all flesh.

When the baboon came out of his preoccupation with commiphoras, the secretary bird had gone, and there was no sign of the dead mongoose. He sat alone again on the open plains. Vultures, wheeling back and forth overhead, were the only visible forms of life.

The vultures were interested in the baboon, whom they understood clearly to be in some form of trouble. They rarely saw a solitary baboon, and almost always the sight presaged death. Slowly, they gathered over him, scores of them drawn to the spectacle by the lack of food opportunities anywhere else at that moment. The baboon, looking up, became apprehensive. He turned back toward the island. He had seen too many vultures tearing corpses to pieces to feel comfortable with them so close. They spoke of death to all creatures.

The ferocious competition for meat among the vultures united them in permanent tension. Two to three hundred birds often gathered around a single lion kill, which meant that all of them were hungry nearly all the time. They were visitors at many feasts but diners at few. It was not merely the competition among the vultures of one kind that was ferocious, but also the bitter battles between the different species of vultures who had separate skills. There were those skilled at ripping hides open and others with a capacity to break up and swallow small bones. A third type scavenged at the fringe of the other squabbling birds, surviving on the smallest scraps of all. A fourth, if left in possession of a really large kill like an elephant or rhino, often entered the big carcass and ate it from the inside.

The sound of vulture wings hissing through the air spoke of unknown, fearful things. During the long days of rain the vultures studded the tops of trees like black fruit, motionless and silent. They were able to go days without food, but when one vulture was hungry, then all were hungry. Scarcely any kill was possible on the plains without their knowledge. Their falling bodies spoke a universal language, as well understood by the other animals as by the vultures themselves. This language was especially understood by one of the large storks who, like the secretary bird, had developed an appetite for meat that rivaled that of the vultures themselves.

When the baboon reached the island late one afternoon, the marabou storks were waiting for him. He skirted the southern shores where they were standing and clambered up the rocks as far from them as possible. Almost everything about the marabous was grotesque. Hunched down, with necks pulled back into their bodies, they looked decapitated from a distance. Close up, grotesquerie became overwhelming. The long black legs were ungainly and daubed with masses of excretal uric acid. When the neck shot up out of the body, it appeared scrawny and ill feathered. The head was spotted with nameless encrustations, or scabs, and led to an oddly efficient-looking beak to be attached to such an incompetent-looking body. In a flourishing finale, the marabou had a long, pendulous sac of skin on his chest which could expand into a startling magenta bulb. A second sac, hidden under white feathers at the bottom of the neck, was sometimes revealed when the marabou raised his neck feathers to display it.

The marabous seemed to be creatures put together in a series of clumsy additions, an evolutionary attempt to emulate the vulture while remaining a stork. When one of the marabous turned into the wind and spread his wings, he became a grotesque flying instrument. The wings were nearly twelve feet broad, planing surfaces which thrashed at the air with incompetent clumsiness until finally this epitome of ugliness caught an updraft and rose at dizzying speed. But a moment later, the marabou decided not to fly after all. He came plummeting out of the sky back to the island, a chunk of flying wreckage apparently out of control, with wings akimbo and feet flung down as if a crash were imminent. A moment before he was due to crash, his pinions rasped and coughed harshly against the air as they sawed the clumsy body to a bounding, buoyant landing.

The marabous were the great sentinels of the plains and woodlands, their knowledge of the position of abundant food supplies passing from one marabou to another so quickly that the birds arrived at their dining places almost as a single group. Unlike vultures, the marabous were practically omnivorous. Nothing faintly edible escaped their notice. They had eaten grasshoppers and crocodiles before coming to the island. They watched vultures rip up the prey while they scavenged the scraps. They waited while lions gnawed at scattered bones, then used their long, spearlike bills to pick up every morsel of meat from inside the vertebrae and joints, from the skull and pelvic bones. Thus, they competed with owls and falcons and buzzards for a share of the small-bird hunting, with jackals for the young of the tiny dik-dik and mongoose, with hyenas for scavenged meat, both fresh and putrid, and with vultures for a share of any kill.

They were one of the great levelers, arbitrating the fate of many populations under them. They had been present at the baboon's great moth and caterpillar feasts, but as was their custom, they had congregated some miles to the north where the caterpillars had swarmed more thickly than around the baboon's island. They had been drawn from the northern plains, from the slopes of the southern hills, from open woodland and lake country. Several thousand of them had concentrated against the caterpillar advance. While the baboon had eaten away steadily, they had gorged all day, latecomers walking among the bloated forms of their comrades.

Such omnivorous appetite was certain punishment for any creature unwise enough to go on a breeding spree. The marabous were always there. The abundant rains of this year had been coupled with unusually severe droughts in the northern regions. This had brought south an aerial flood of small finches—queleas—in spontaneous migration to the northern fringes of the plains. They had arrived as the marabous were in the middle of their caterpillar feast, and the big birds did not react immediately to the newcomers, which gave the queleas a chance to begin breeding without hindrance. The long, lank grass made ideal nest-building material—indeed, the birds could not come into breeding condition unless this grass were available—and within a dozen days the queleas had spread themselves over more than twenty square miles. Their nests were numbered in the hundreds of thousands. They clustered in low bushes, a gray-green crop multiplying itself daily, the breeding so rapid that the quelea horde soon possessed eggs, fresh-hatched nestlings, and fledglings simultaneously.

As the caterpillar feast waned, the marabous discovered the queleas. They harvested the finches as methodically as they had the caterpillars. The marabous did not care whether the nests contained brooding females on eggs, which they ate, or young nestlings, which they ate, or a scattering of fledglings, which they also ate. They pursued the queleas with a solemn, stalking dignity, picking up one scrap of life at a time.

Eventually, two thousand marabous had gathered at the quelea metropolis. The ongoing feast lasted nearly forty days, with many fluctuations as the queleas' numbers were temporarily suppressed, only to rise again. The marabous destroyed twenty to thirty million eggs, nestlings, and adults, but the moment a nest was destroyed, a new one was constructed elsewhere. The millions of queleas had an overwhelming advantage over two thousand marabous, and despite the great losses—magnified by countless smaller hunters—they completed their breeding and raised their numbers from five million to more than forty million. Then they moved off in a series of dense, streaming flocks which looked like black clouds of flies. The marabous remained, stupefied with food, so full of flesh and eggs that few could fly. The hyenas came on them and killed many of them before digestion was half-completed, and the marabous took off, clattering their beaks in anger.

If the baboon could have been present at such a feast, he would have exploited it to absolute surfeit, perhaps even vomiting up his giant meal before going on to eat more queleas. Twice before in his life he had encountered quelea hordes, and each time the baboon troop had spread wide to gorge. But the queleas now were many miles away from his circumscribed territory. Still, he continued to move amid a flood of birds stimulated by the aftermath of the rains. He pirated larks' nests in thick clusters of grass, chased fledglings through the trees of the island, and stalked ground-walking birds when he saw them trying to hide themselves in scanty cover on the plains. He moved among a decoration of birds who occupied their places on the plains with a perfect and enduring precision. The queleas only appeared after heavy rains. Other birds would be seen when drought began. The flappet larks took to the air and rapped out their mysterious messages with horny wings and seemed to inspire a large grasshopper to hover in the same way and to stutter out a similar message. The survival of the birds was balanced against the contrasts of drought and flood, of famine and surfeit. Each adjustment made sense, whether it was the touching display of the widow birds, the sensitive skill of the larks in concealing their nests, the fecundity of the spur fowls and guinea fowls, or the caution of geese sitting silently in the middle of ephemeral pools. Only the ostrich seemed to confound the rules of reason, hiding the secret of its survival behind a facade of the bizarre.

As the baboon watched, an ostrich created a nest within sight of the island, placing it in seeming ignorance in the center of a hyena roadway, marked with white droppings. The ostrich managed to lay six giant eggs before the inevitable night collision, and in the morning, the baboon saw only small vultures picking scraps of yolk from the shattered shells.

Ostrich nests might contain the eggs of one ostrich or of a dozen. The egg laying might be leisurely, spread over days, or it might be quick, completed in a few hours as female after female added her share to the communal nest. With up to sixty of the huge eggs in the bowl of a nest, one bird could not hope to cover them all. A sixty-egg nest might produce only two youngsters, but conversely, a ten-egg nest might produce a full hatching. The ostrich nests were a magnet, drawing hunters and game lovers. For the baboon, the great eggs were a constant puzzle. He relished their contents, but he was unable to get inside the globes. Unlike the chimpanzee, he had never mastered the art of tool using, so he had no way of smashing the egg. But one of the vultures used a small stone to rap the shells open. The baboon watched these vultures at work, and when they were successful, he drove them off.

Lions moved toward an incubating male ostrich while he waited, head raised high and fearless from his absurdly exposed position in the middle of the flat, empty plain. At the last moment, when he was looking directly into the eyes of a charging lioness, he leaped up, but not quickly enough to avoid a stinging blow on his rump. The lioness shook a thick black clump of feathers from her paw and turned to the nest.

For an hour, the lions played at the ostrich nest, batting the eggs about with contented grunts and groans and belches. They picked them up between forepaws and rolled them over, crunching other eggs beneath them and spreading thick yolk over grass and tawny hide. The lions were not content until they had shattered all the eggs. In their wake came the more cautious hyenas, egg lovers who patiently licked the yolk from the grass and from the broken shards of shells. After the hyenas came ants and beetles and other creeping creatures who removed every scrap of matter from the shells.

Once the egg hatching phase was completed, the defense

of the ostriches seemed more sensible. Their greatest enemy, the hyena, favored young ostriches, but when she tried to hunt one, she found herself faced by a creature who towered above her and was equipped with the most powerful kicking legs on the plains. These legs, coupled with a speed that was greater even than the fleet cheetah for short bursts, made the hyena herself vulnerable. She was in danger of being kicked high into the air, her rib cage crunched or her back broken.

Like the giraffe, the ostrich was difficult to attack. If the hyena seized a leg with one of her bone-crushing attacks, she would almost certainly get kicked by the other foot. She could not jump and grab for the vulnerable neck or head because they were too high. And the body of the bird, with its great mass of thick black feathers, gave her no encouragement. What would the hyena bite for when she could not determine which was feather and which was flesh?

The young ostriches came under a kind of community guardianship which did not seem to be based on the cooperation of the adults. Twenty to forty youngsters, all of them three to four months old, followed a single hen and cock ostrich, but occasionally—and this was more difficult to explain—huge collections of a hundred or more young ostriches followed a single pair of adults.

One bright clear morning, nearly three hundred ostriches appeared at the baboon's island, all of them youngsters except for three adults, all of them welded into a single group which must have been a coalition of a score or more of nests. The ostriches paused at the island, then headed steadily into the sunrise, showing a perfect silhouette, a frieze of creatures strung along the red horizon, taking their secrets with them.

Their triumphant survival in a world where lions hunted

them for fun, where jackals pestered their young, where hyenas listened to the tapping of beaks on the insides of eggs and awaited the hatching, where both vultures and mongooses had learned to use stones to bash their way into the giant eggs, was another mystery in a place where the rhythms of life had not changed for hundreds of millennia.

The ostriches had gone, and vultures wheeled away into the blue infinity of the moving day. Secretary birds perched on flat treetops near the baboon, and spur fowls hid themselves. Larks sat on their nests while finches flitted in bright foliage and brilliant sunbirds and bee eaters were briefly visible in the hot afternoon. Then dusk came quickly to the island which was filled with the silent forms of great and small owls and echoed to the cheerful voices of nightjars coursing across a sky made purple by the silent advance of night.

XI

The big-eared dogs appeared like ghosts strung out along the back of a low ridge in the smoking light of late afternoon. They were seen immediately by every animal in the vicinity. Behind them, the radiated light of the setting sun writhed and twisted and seemed to make the motionless bodies of the dogs move too. Their dominating position on the ridge near the western island made them able to look down on the animals dispersed before them. They saw lions sprawled lazily in the distance with a solitary jackal nearby, and a great concourse of zebras and antelopes spread to the horizon.

The grazing animals brought their heads up sharply, and

a multitude of eyes looked at the dogs. No animal could remain neutral when the dogs were present, not even the other hunters. The gazelles and the wildebeests had an almost mystic capacity to know whether or not the dogs were hunting, but the zebras seemed to have none of this capability and showed less concern when the dogs were near, even though the dogs sometimes killed them.

The sky turned yellow and then orange as the sinking sun thrust its light through curtains of cloud. The dogs did not move, and none of the watching grazers panicked; the vital danger signals were absent. No animal truly understood the dogs, except to know when they were dangerous and when they were not. For the rest, their lives were an enigma; they were periodic visitors, periodic disseminators of terror and disaster, and that was all.

The prey animals understood that their chances of survival in any attack were many hundreds of times greater than their chances of being killed. And it was entirely possible that the dogs would not make an attack at all. Many of the grazing animals returned to cropping grass, assuming that the dog threat would resolve itself.

The dogs stood in a long, straggling line on the smoking skyline, their blotched bodies still as stone. If they seemed inscrutable, it was because the organization of their group was one of the most complex on the plains, an odd mixture of authority and camaraderie, more difficult to define than that of the hyenas because a hyena could always split away to take independent action, as the old female had demonstrated, while the dogs were totally dependent on their group organization, their group ethic. Without it, they were dead. They had created a cooperative society in which each dog was expected to be able to hunt, to defend the den, to feed the youngsters and guards. They had abrogated the age-old custom of the meat eaters which demanded that the strongest and the most aggressive eat first.

The leader appeared to be a large male, his body bulkier than the others, his stance more tense, more dominating, but the hierarchy of the dogs could not be measured by size or sex or any of the other simple externals. The true leader of the group was a smaller female who stood at the end of the line of animals and who was now the first to move, licking one of her forepaws as though she had lost interest in the animals before her. The large male simply dominated the other males, who like male hyenas, were sometimes separated from the female organization.

The large male had led the group to the ridge, but that did not mean he would decide whether to make an attacking run. He might lead the dogs toward their victims, and he might, indeed, make the kill, but that did not define him as the leader. Contradictory and puzzling, the dogs were the only larger animals of the plains to have developed such finely honed cooperation.

Now, they were undecided because they were exploring the possibilities of moving into new territory. Their territorial arrangements were not like those of other animals and could not be described in terms of the normal limits of land, or of the relationship of the pack to other groups of dogs or to any other animal. Their territory changed according to mysterious needs that sent them ranging from one end of the plains to the other. They traveled from the foothills in the south to the edges of the woodlands.

If the dogs seemed unpredictable, it was perhaps because each animal varied so much in personality. The large male had achieved one kind of dominance by his aggressiveness, by his single-minded need to follow each attempted kill to the limits of his physical endurance, but the dog who stood next to him, a brindled, yellowish animal who seemed light-headed by comparison, was perhaps more intelligent in that he quickly knew when a hunt was futile and frequently stopped running seriously only seconds after it had begun.

While the lead dog sat watching other dogs play in times of leisure, the brindled dog gamboled among younger dogs as though he were one himself. In contrast to the lead dog's singleness of purpose, he was impulsive, and frequently branched away from the main pack to hunt alone, a breach of the group contract which brought him into conflict with the bigger male.

The variations in personality among the dogs ran down the hierarchical chain until the lowest dog was reached. He was the runt, the smallest and weakest of a large litter of eighteen pups born three years before. He was the slowest runner in the pack, and his survival was expressed not by his body but by his brain, where lurked a determination none of the other dogs possessed. He arrived last but he immediately moved to be as close as possible to the dominant dogs, as if he found the strength he lacked through the association. He stood to one side of the dominant male at this moment as the antelopes continued grazing, most of them certain now that there would be no dog attack.

The sun had gone below the horizon and the changing light deepened into a rich crimson glow that turned the dogs' eyes red and washed their mottled coats with pink. The female at the end of the line moved toward the center until she came to the dominant male. She licked him on the neck and he turned as she nipped his left foot. He braced himself but she pushed with her body and he stepped aside, moving slightly backward so that the bitch now stood ahead of all the dogs. Then a second female who had been standing near the end of the dog line moved toward the center, and she, in turn, licked, nudged, nipped, and pushed aside the dominant male so that she could stand next to the lead female. A ripple of anticipation moved down the line of dogs; tails wagged, mouths opened in anticipatory panting. One of the youngest dogs gamboled for a second, but he was growled into stillness by an older dog.

Because the dogs were governed by their group needs, all of them were conscious of the distance that now separated them from the cluster of dens they had established a score of miles to the east. Caring for the fifteen juveniles not yet ready to accompany the pack were two mothers and five guards. These dogs and puppies depended on the success of the hunting pack for their food. The twenty-one dogs in the attack group must kill for more than forty, then transport the meat back to camp.

The smoking crimson light turned purple, and dark clouds moved overhead in the grip of a strong breeze. It was not going to be a clear and moonlit night. The encroaching gloom set the moment of decision. The lead female started down the ridge, the other dogs streaming quietly after her, their feet scuffing on the grass. The attack was to be made frontally and was to be unexpected because the gloom had blotted out the waiting antelopes and zebras. The dogs ran with the silent discipline of their kind.

They were the most dispassionate of the hunters, since they possessed an expertise which did not depend on strength or cunning or speed. They ran down their quarry in relentless, unhurried pursuit that seemed to lack all feeling. Yet they needed their moments of fun, both as pups and adults, and in their play they were unique among the hunters.

They delighted in baiting the big and the strong. Perhaps they were testing for weakness, since they harassed elephants and rhinos as well as smaller animals, but it seemed more like play. Before their last den move, the pack had found an irascible rhino. They circled the angry animal and some of them frolicked together affectionately around him, nipping each other's flanks and licking faces. The rhino had charged at the brindled dog and they had all scattered. He charged them again but by then they were behind him, nipping at his hindquarters and his ridiculous tail. The bites did not hurt, but he whirled to see where they had come from. No dogs. They were behind him again. The runt yipped with pleasure at this sport. After several hours of the baiting, the rhino was exhausted, his dignity shattered, and his protruding tongue coated with eager flies. His wheezing gasps and piggy squeals mocked his armored skin which had been useless against the deft provocation of the dogs.

The lead female had turned away from the defeated rhino and the other dogs followed her. Later, the pack found two elephants who were drinking at a water hole. They yipped in delight. Here was a new game, but they played it according to the same rules: encirclement, mock charges, nips at towering back legs. The elephants took the teasing seriously. They trumpeted acute concern, their trunks raised and ears flapping, all thought of water gone as they raced away. Well content with their provocative games, the dogs streamed off on their own endless safari. If this were indeed play, it represented humor of an odd sort. The dogs seemed to have some intuitive knowledge of the weakness in the animals they tormented. Frequently, the play was tinged with cruelty and became a symbol for something dark in their natures.

There were times when they misjudged their victim. Once they had come across an old lion, long since deprived of his pride and holding onto his dignity with difficulty, and had silently surrounded him. He had risen with a groan. He was not scared by dogs, but they had been a lifelong nuisance. Perhaps he felt that one quick charge and some really heavy roaring might drive them off. But the dogs had yipped their delight at these tactics, and suddenly they had all begun to behave like puppies. A gavotte of dogs had risen on hind legs, pirouetted, barked, leaped, encircling the lion again and again. Despite his age, he had never seen such an extraordinary performance. The dancing became more intense as dogs licked muzzles, nipped playfully, and rolled over gripped in affectionate embraces. It was as though they had forgotten the lion existed, but whatever motivated the dogs had not touched him, and he charged once more. This time, he knocked over the brindled dog with a swipe to the hindquarters that took off a pawful of hide, and sent a female spinning a score of feet. The playing stopped abruptly. The dogs stood frozen while the two injured animals whimpered softly. Then, all the dogs howled together, a melancholy and reproving sound, before they disappeared.

The dogs were not as big as hyenas; yet their organized attacks inspired a special kind of terror in the grazing animals. A week earlier, the lead male had found himself baffled by the agile zigzagging of the gazelle he was tracking, and he had been forced to stop, panting hard, until three of the other dogs had hastened over to him and the hunt had continued combining their strengths. The victim had been pressed hard by one dog, then by another, so that the desperate edge of its energy was soon gone, its zigzagging no longer so skilled. The dodging animal had blundered into a flanking dog and was killed by him, eliminating the need for a communal attack.

The gazelles always ran as soon as they saw the approaching dogs, but the wildebeests presented a different problem. They often stood their ground, the bulls to the fore of the herd, the cows and calves behind. In one recent encounter, the charge of the dogs had brought the immediate response of a wildebeest countercharge, which was a rather desperate affair with the defiant animals holding their ungainly heads well down, their long hair streaming and the entire herd tense as spring steel.

The dogs had easily avoided the charge and then spread out disconcertingly wide, making the wildebeests vulnerable indeed. The attack became more psychological than tactical because the dogs did not dare direct another attack into a mass of angry bulls, and so as long as the wildebeests stuck together, they were safe. But the dogs understood the nature of the individual weakness lurking in the wildebeest herd, and they sought to expose it. They feigned charges and fled from the bulls, then turned and flanked them until the apprehension of some animals within the herd increased unbearably.

The animal they had finally killed was young, as they frequently were. He had become jittery even before the hunting dogs had attacked. He had broken away from his mother, rushing past the bulls who might have saved him to head for the illusory security of the open horizon. His mother followed him and the two animals hurtled toward the dogs who wheeled responsively wide of them. There was no need to drag down a calf so near to the bulls. The strategy was to fragment the herd. The big male dog led a feinting charge against the defending bulls while the dominant female ran at the flank of the fleeing calf. The wildebeest mother quickly realized how hopeless the position was and turned back toward the herd. Then, all the wildebeests broke, stampeding in the opposite direction so that the calf was quickly isolated by his cluster of killers. The mother, already well buried in the fleeing wildebeest herd, never saw the first bite which tripped her calf, or the second which ripped his body, or indeed any part of the bloody minutes that followed in which the calf was transferred into the stomachs of the dogs to be transported at forty miles an hour back to the den, and so into the stomachs of other young animals and their guardians.

Now, with night falling swiftly upon the plains, the surprise of the dogs' attack was complete. Because the grazers themselves hunted at fixed times every day—always for one hour after sunrise and for one hour at sunset when their prey was most intent on feeding—they had assumed that night would make them safe from the dogs. When the silent, streaming pack cut among them, panic spread, and the grazers spurted away in flight from the invisible dogs loping through the deepening gloom.

At first, the dogs remained fairly close together in a long narrow attack formation, the dominant dogs running ahead, but after a few minutes, they began to spread out, thus widening the area of their reach. Zebras and wildebeests, surprised by the night appearance of the dogs, wheeled away to either side, only to find themselves in the track of other dogs, which sent them hammering off in yet a different direction where still more dogs appeared. It soon became apparent that this technique of attack was deliberate. The dogs could not see each other at day distances, and thus were unable to divine the most vulnerable animal to kill, so they were using the strategy of surprise. It would only be a matter of time before an animal wheeling away from one dog, succumbed to the next.

There was an undertone of desperation to the dog attack. By the group ethic of the pack, the night hunt was dangerous to the point of foolhardiness. High-speed running in the dark over uneven or unknown ground was fraught with hazard, but the group hunger, spurred more by the distant imperative of hungry pups than by their own empty bellies, forced them to this attack, forced them to try for a large animal.

The fifteen young dogs left behind in their camp were favored creatures, the first to eat from any hunt. Not for them the long vigils of starvation endured by hyena youngsters when their mothers went hunting, or the hunger of cubs when the lion pride could not migrate to follow its supplies of meat. Instead, they had the almost constant attendance of mothers who were so maternal that they sometimes tried to steal each other's pups to nurse.

The guard dogs left behind by the hunting pack patroled the den and were so fierce in its defense that they tried to keep prowling lions and hyenas at least one thousand feet away. It was this ultrasensitive concern that had created the present, and growing, hunger of the dogs. In their previous camp, where the cubs had been born, lions had appeared when the eldest of the young dogs were forty days old and active enough to play outside the den. The two main guard dogs had been called away to hunt, but the young dogs had heard the lions in time and had disappeared into the underground dens. The lions padded around the entrances, thrusting their big faces into the holes and probing with clumsy swipes of their paws.

The dogs had needed no other warning. As soon as the lions had gone, the females had emerged from the dens and run off. They had returned with the other dogs and immediately the entire group had moved, the oldest youngsters chivvied along by the dogs while the youngest were dragged in their mothers' mouths for maximum speed. They had moved to a new camp site a score of miles away, where they quickly enlarged an old aardvark burrow, working with smooth cooperation, but the move had meant five days without food, which the dogs found uncomfortable.

By spreading out on the night plains, the dogs were causing major changes in the disposition of the animals. A broad, streaming flight to the east united thousands of zebras and wildebeests. The noise was so great that the baboon wakened on his island and listened apprehensively. The leopard, who was halfway across the dangerous territory between the two islands, turned toward the sound and sniffed the night air. He ran a dozen hesitant steps, his cat brain filled with the darkest premonitions. But steadily, and without pause, the noise, like the sound of low thunder, grew louder.

The long thrust of the dogs' attack smashed the sleeping formations of the grazing animals into thousands of meaningless units as gazelles ran among zebras, and wildebeests fragmented. The panic of the grazers contrasted with the steady pace of the running dogs, who showed no excitement and uttered not a sound.

Normally, their victims ran in circles, which gave the dogs an advantage; they simply cut a straight line across the inside of the circle, rapidly decreasing their distance from the prey without too much effort. But now, the attack was frontal, and the movement of the animals linear, at least for those who stayed ahead of the dogs.

On this night of the hunting-dog attack, two jackals found themselves caught in the fan-shaped area disturbed by the running dogs. Like the leopard, they understood the danger. The sound of the galloping zebras and wildebeests, the wild cries piercing the night, the inquiring hoots of hyenas in the distance all signified drastic changes in the world where they had been hunting. During the day, they usually hunted together, spread wide apart but keeping in contact with short, characteristic barks. They followed hyenas and nomad lions in the hope of scraps, but their favorite animal was the cheetah, who was so selective a feeder that he left much of his kill for them to scavenge. Now, however, comradeship dissolved in panic. The vixen jackal ran directly east, hoping to find refuge on the island of the baboon and leopard where a small cave among the rocks had given her shelter before. The dog jackal hesitated, bolder than the vixen, but not more knowledgeable, hoping that the flight of the grazing animals could be turned to his advantage. While he waited, the first of the fleeing zebras came upon him, and then he was running for his life.

The jackal was an insignificant scrap of fur and guts lost in a melee of flying hooves and chopped shards of grass. It did not matter that he darted one way, then another, the flight front of the zebras was too wide for him to run beyond. But he remembered a burrow he had passed in previous days, one which he had not investigated to see whether it had an occupant. He headed for this burrow.

The zebras ran straight, tending to compact into groups, and this made the jackal's flight easier. He judged the position of the burrow accurately so that he came upon it while he was still running between two galloping zebras. In his haste, he had no chance to sense the danger of the burrow. He charged full tilt into the hole and disappeared.

It was the jackal's bad luck that the burrow, a converted termite city, was occupied by a large python curled protectively around nearly one hundred eggs. His explosive entry brought an immediate response from the python. His face was filled instantly with great curved teeth that drove into the upper part of his muzzle and brought from him a scream of agony. He tried to pull back but the snake's grip held him fast. Because she was protecting her eggs, she was incapable of seizing him with her body and crushing him, so she hung on with her teeth, shaking him as though she, too, were a dog. He screamed again. Two dogs passing overhead heard the scream, but it meant nothing and they ran on silently.

The jackal had learned that to turn and face an enemy sometimes stopped the attack, so now, he attacked, throwing himself forward so that his muzzle slipped off the curving teeth. He seized the python just behind her head. The burrow erupted in a mad confusion of snake and jackal bodies. The eggs were kicked and smashed to pieces. Young snakes, several score of which had been about to hatch, wriggled helplessly between the writhing bodies. Some were kicked to the surface where they would be picked up by eagles, marabou storks, and secretary birds in the morning.

Free of that terrible grip, the jackal faced the secondary danger of being constricted. He lacked room to maneuver in the burrow and could only dart back and forth in the hope that he would not be seized. The python gave up first. She was acutely uncomfortable in such a small space and she flowed out of the burrow into the darkness. The jackal fled moments behind her, sniffing the now-quiet air. Both zebras and dogs had gone to the east. He trotted off.

The dogs had never increased their collective speed, and they made their kill out of a kind of inevitability. No horse or antelope was evolved to run indefinitely, so the chase was merely a matter of staying with the fleeing animals until one of them weakened. A pregnant mare found herself flanked by two of the running dogs, but she was too exhausted to swerve, or increase speed, or stop, or do anything. Her family unit had been broken earlier. Her stallion was far behind her, outrun by the dogs himself and helplessly watching as his mares scattered to either side and his juveniles were fragmented and lost. The mare scarcely felt the first lunge of the dogs, a bite that was accurately aimed at the softer, less taut flesh between her thigh and stomach. She was barely conscious that a dog hung onto her there, but she did feel the second bite which caught one leg just above the knee joint. She elected to stop, kicking in an attempt to dislodge her attacker. And in that second, her bad judgment was revealed when a dozen dogs came streaming silently out of the night. The mare could do nothing as she went down under a swarm of eager jaws reaching around her hindquarters and up her neck. Numbed, her eyes glazed as she was devoured.

The dogs left the killing scene where pieces of the zebra's body lay scattered about among the bones, their stomachs filled with great chunks of unprocessed food. Now, they sped east at half the speed of the cheetah's chase. They passed the baboon's island, crossed the leopard's frightened scent, and ran until they came to their camp.

The guard dogs came out to meet them immediately, and the pups bounded forward hungrily. The hunting dogs began disgorging large, solid chunks of undigested meat. The pups clamored at their muzzles and many took the meat before it could fall to the ground.

Some chunks were so large that no pup could eat them. These were eaten again, sometimes by dogs other than the ones who had disgorged them, and then they were regurgitated once more. The largest pieces were eaten a third time so that the meat was reduced finally to a form suitable for young puppies.

Home at last and their camp secure, the dogs revealed themselves as members of a family that was like no other on the plains. It was an egalitarian family, or nearly so, only the mothers possessing special status. They might not greet the other dogs as intensively as they were greeted, or play with the same kind of abandon, but if they were slightly aloof, the others were not. They seemed tied together by a mass of rituals that were performed so enthusiastically and so repetitively that they had become a part of the enjoyment of the pack as a unit.

Some dogs licked the udders of the females in greeting them, as though they had become pups again and were supplicating milk. Some females crept underneath the males as though they were nursing pups returning to *their* mothers. The young became hungrier as the rituals and demonstrations continued, and pushed closer and closer to the adults until the food was invisible as it passed from mouth to mouth. The adult dogs importuned each other and those who had not gone out on the hunt were fed, just as the pups had been, to the accompaniment of nipping at jowls and yipping for the meal which the communal ethic told them was at least partially theirs.

XII

The hyraxes had been reduced to a shadow of their old numbers and the grass grazed to the height of a mouse's ear; the birds' nests had been pirated and the water hole drunk to a viscous trickle. Still, the baboon hung on through long, frustrating days and nights of terror, surviving at the edge of desperation on the scanty food supplies he eked from the hostile world around him. But he faced an inevitable decision: he would have to move, and his head understood this as well as his aching and empty gut.

He sat on the top of the highest rock of the island scanning the northern horizon. He watched a thickening of vultures concentrated over a slow-moving herd of gazelles. He saw wildebeests in the foreground and lines of zebras in the east, but there was no sign anywhere of lion or hyena, jackal or dog. The leopard's tree sat empty in a splash of golden sun. Now was the time, now. Yet still he hesitated. Memory, although ephemeral, was sometimes so deeply driven into the unconscious of an animal that it was capable of inhibiting action. To the moment of his death, the baboon must remain conservative, touched by memory or by instinct or by whatever it was that always assumed the worst.

When he finally steeled himself to leave the island, it was with the same exaggerated caution he had displayed on all his other forays. Watchful and wary, he carefully clambered down the rocks through overhanging shrubs, then went slowly down the final, grassed slope to the plains themselves. His black, half-upright form gradually diminished to become an insignificance of matter lost in the vastness of the plains, and before the afternoon turned downward toward night, it had disappeared.

The baboon moved among grazing animals. Nearly all the females were pregnant, and the gazelles concentrated north of the island were excited and jittery. The pregnant does stood a short distance from their herds where the grass was long enough to give some shelter to the fawns who were about to be born. The gazelle bucks seemed to have lost control of their does at this moment of birth. As each newborn gazelle dropped from its mother's body, the doe kept turning her head to all horizons and to the sky itself for the first sign of an enemy. The new mothers were aware of their fawns' vulnerability to hunters as small as jackals and as unexpected as the larger eagles. Even small birds were abruptly charged by the nervous does if they alighted nearby, and the bustards found they could no longer walk among the gazelles without being charged so repeatedly that they eventually flew off in disgust. The does seemed to understand that they were safer together; they united in nursery herds while the other gazelles grazed a short way off.

In a solitary tree, a huddled black form sat frozen with indecision, a reluctant witness to the birth time of the gazelles. The baboon had come as far as courage and light allowed, and now he awaited the signal for his next move.

The gazelles were prime victims, and they responded to their many enemies with fast breeding. They held their young in gestation for five months, and unlike most of the other grazing animals, they mated and fawned throughout the year, although their births peaked somewhat during their time on the plains. Of necessity, their fawns were precocious, independent after only forty days of life outside the womb, ready to bear their own fawns by the end of the first year, ready to conceive again a scant twenty or thirty days after giving birth, mating and breeding in what appeared to be a constant race against death.

The baboon watched youngster after youngster stagger to its feet only moments after birth, making pathetic efforts to walk. He felt the baboon troop ethic stirring inside him; this was an unforgettable opportunity to hunt, but he had never done it alone, and he hesitated. In their efforts to walk to hiding places, the gazelle youngsters were trying to win their own game with death, trying to reduce the chances of being cut down by the hunters gathering around them and above them. They lurched away under the wheeling black crosses of the vultures, stumbled into hollows and long grass where they held themselves absolutely motionless, gripped by orders given them from the deep evolutionary past.

Their mothers circled the staggering young, suckled them, licked them. The first rule of survival among the breeding gazelles was that no hidden fawn should ever be betrayed by its scent. The fawns had skin glands which did not work at first and so gave off no aroma, but other scents made them vulnerable. The mothers ate all the body wastes to eliminate telltale odor. The baboon looked down on a fawn hidden in grass at the feet of a jackal, as keen-nosed as any of its kind, but the jackal neither saw nor smelled the fawn.

Jackals had gathered all along the southern fringes of the fawning ground and they were trying to spot suitable victims in readiness for quick, rushing attacks. Simultaneously, each mother put herself between her fawn and the nearest potential attacker. The moment a jackal got into his attack stance near the baboon's tree, a doe rushed forward, making a series of short, stabbing charges designed to confuse rather than to harm the jackal. When the jackal was not deterred by these feints, a second gazelle mother raced forward to help in the defense of the fawn, and the jackal slunk away.

Elsewhere, though, other doe gazelles were not coping so well. Jackals spotted concealed fawns whose mothers were grazing and darted toward them. The rush of the jackals brought from the fawns characteristic distress cries which alerted their mothers, who rushed to help. One doe knocked down a jackal and some of the jackals were intimidated and wheeled away. But others stood fast, driving back the gazelles. When the way was clear, they seized the fawns and strangled them before eating them. The helpless mothers watched the feast for some moments, but then they turned away and began grazing within sight and sound of the fawns being eaten.

The buck gazelle alternated between concern and agony during the fawning and the attacks of the jackals. The distress cries of the fawns agitated him. He watched the thrust and counterthrust of does and jackals, but he did not join in the defense. He stood uncertainly, twitching his short tail. The jackals were still some distance from his territory. One of his does charged a jackal who fled toward the buck, spurring him into action. For a moment, it appeared that he was charging the jackal in mutual defense of the fawn, but his charge hit the doe, shouldering her aside. The doe stumbled and turned away, her life saved perhaps by the buck's charge, but her fawn strangled and eaten.

Both gazelles stood near the jackal, tails wagging furiously. The jackal's cunning eyes watched them from above his crunching jaws. The tableau appeared to demonstrate a natural truth—the doe saved, the fawn sacrificed— except that the buck had not, in fact, saved the doe at all, nor did he, indeed, have any interest in the fawn. He had no mechanism, no equipment, no stimulus to save fawns. His concern remained perennially that of territory, and he would have done anything in his power to prevent one of his does from leaving it, which the female, in pursuit of the jackal, had been in danger of doing. Both animals, the one imbued with what appeared to be parental love and the other having none of it, had rushed forward to fulfil their predestined roles.

They were mechanical creatures, mechanically stimulated, mindless perhaps, but after the jackals had gone, the mothers who had saved their fawns or had not been required to protect them moved cautiously into the refuge territory. They all remembered the positions of their hidden fawns, and one of them called softly. The fawn immediately jumped up and ran to her, gamboling as it came, then settled to nurse. When this was done, the mother watched while the fawn went off alone and concealed itself again. Having memorized its hiding place, the mother was free to resume her grazing. The buck ignored the does and fawns, and with the jackals gone, he grazed the short green grass before him as though nothing had happened.

Meanwhile, south of the gazelles, a shattered herd of wildebeests was slowly gathering itself into order again. The black bull's herd, always nervous at this time, had suffered an almost total loss of integrity during the ferocious and relentless attack of the hunting dogs. Normally, the herd ran together in such attacks, knowing through experience that it was harder to cut single animals from a large group than from a small. But the simultaneous appearance of dogs across a wide front had disrupted this habit, and the entire herd—more than twenty thousand animals—had disintegrated. The black bull had found himself fleeing in the opposite direction from his family because five fast-running dogs had interposed themselves between him and his cows.

He had tried, as was his custom, to confront them; he had even charged blindly at a dusky figure, only to discover that it was a solitary hyena, herself caught in the maddened confusion of the night. He had attempted to change the direction of his flight, or at least to resist the pressure on him to run to the east, but the moment he turned south to follow his family, more dogs appeared and he wheeled away again to the east. By this time, he was so far behind the fleeing wildebeests that he could not even hear their hooves. No other bulls were standing their ground. In fact, at that dark moment, the night air silent around him, he divined that he was alone. When hyenas appeared, he ran again, this time blindly because he had no notion of where he should go, no imperative to defend anything. He was aimless, hopeless, helpless.

The hyenas ran with him steadily, in much the same hunting rhythm as had been used by the dogs. Like the dogs, they tended to turn their victim in a circle, and the bull could not know that he was being forced steadily toward the south in a great, sweeping arc. All he knew in that mad run was that six hyenas were at his flanks and withers and he felt the sharp sting of a bite, the dragging weight of a body hanging on to him, followed by another sting and another weight. At that moment, he dragged himself into a crush of wildebeests so thick that he was knocked down and trampled underfoot by animals running in every direction.

The hyenas were trampled too, and to escape they had to relinquish their deadly grips on the wildebeest bull. Getting to their feet, they found themselves lost in such a crush of hysterical animals that they were incapable of functioning as hunters. They lurched this way and that; some were knocked down again and trampled and the screams of one of them, crippled by a wildebeest, started a new panic.

The black bull was separated from his family, but that was the least serious result of the hunting-dog attack. Unlike the gazelles, the wildebeests were adapted to calve in as brief a time as possible, a device that cut short the period of birth vulnerability. The cows were able to control the moment of birth, sometimes delaying it for several days, but the dog attack had brought many of them prematurely to delivery. At dawn, the black bull found himself standing on a calving ground with the first of the calves appearing even before the sun rose.

The calves were light, fawn-colored creatures compared with their almost black parents. The calving continued through most of the morning, the cows gathered so densely in one place that few hunters were able to drive them from the newborn youngsters who were standing and suckling within minutes of their births.

Because the cows had such precise control over the moment of calving, the choice of this time presupposed that all danger was past. This was true in the case of the dogs. But the confusion of the dog attack had alerted the hyenas, who were drawn from the edge of the northern woodlands by the cries of the hunted animals.

And so, the black bull, still searching for an aromatic clue to the whereabouts of his family, became embroiled in yet another attack. The hyenas came on the packed wildebeests in a series of circling movements that compacted the calving animals, but the hammering feet, the whinnyings and bleatings, the roarings and whoopings, the sudden appearance and disappearance of hyenas dashing madly from one kill to another unnerved the cows. When two groups of hyenas united briefly to circle several hundred cows and calves standing near the black bull, the cows were close to panic. The hyenas, wary of the massed wildebeests, did not press an attack that might have been successful. It took another hunter to complete the disaster of the wildebeest morning.

The old lion, lying half-asleep among some tussocky grasses, heard the wildebeest uproar in the distance. He listened for a while, understanding the nature of the attack, but was disinclined to join in exploiting it. As the noise continued far beyond its anticipated span, however, he finally got up and limped toward it. When the wildebeests were really badly panicked, he knew, they sometimes left calves behind as they fled, and these youngsters provided easy kills. The calves, having nowhere to go, could be eaten at leisure for days afterward. Once, in his experience, he had come upon some newborn calves who had not even had a chance to suckle. They had followed him, a surrogate mother, as he went about the job of killing and eating them.

He arrived at the disaster zone just at that critical moment when the cows were stressed to the maximum by their resistance to the hyenas. He lacked the hesitation of the hyenas about hunting among such a concentrated pack of animals and walked straight toward them. They, in turn, had none of the resistance toward him that they had shown the hyenas. They might charge hyenas, shoulder jackals out of the way, and confront hunting dogs, but no wildebeest could ever face a charging lion. The assembled cows quivered. The calves bleated. The hyenas came closer. In one great, explosive moment, the cows fled.

The flight was so quick, so thoroughly panic-stricken that many calves were trampled underfoot and killed. Calves taking their first drink had the teats wrenched from their mouths. Dust and clods of riven grass struck the abandoned youngsters. The lion, transfixed by the speed of the wildebeest flight, knew now he would feed with the hyenas. When the dust and tumult subsided, a combined cry rose from the deserted calves. Those who had been born earlier ran dementedly back and forth. Some of them charged at the hyenas and were brought down. One ran at the lion and he batted it with a leisurely blow of his good paw. The sun stood hot in the midday sky, and the vultures

227

stacked themselves above the wildebeest disaster. The rumble of hooves gave way to the rending of flesh, the cries of the abandoned, and the grumblings of the old lion feeding full for the first time in days.

The outpouring of calves and fawns created a confusion in which the circumspect baboon moved rapidly, his hunched black form scuttling across the vastness with sustained determination. Witness to gazelle deaths and wildebeest flights and the rapid passage of the hunters, he threaded his way through an invisible and formless labyrinth of danger, measuring the smell of the air, listening intently to every distant sound, watching the circling vultures overhead. The deaths of the youngsters had favored him, and one quick rush from his refuge tree had given him the hindquarters of a newborn gazelle, more food than he had eaten for over thirty days.

Death, chaos, the frenzy of fleeing feet, all were familiar enough to him, and he, like most other animals, understood that many youngsters should not be allowed to survive. This knowledge was as pitiless as the hunt itself. A lioness, watching her cubs play in the sun, reached forward and crunched flat the skull of the smallest. A young elephant born in the shade of trees on the shores of a lake, rose to his feet to walk and was astonished to find himself struck viciously by his mother's trunk when he attempted to suckle. Squealing his incredulity, he found himself charged and driven helter-skelter into the shade of the trees. There, still squealing, he was tusked to death. This act of murder was completed with such a lack of ceremony that it was clearly preordained. The young elephant had been selected as a nonsurvivor, and nothing could have saved him.

For many of the animals, the murder of their young

appeared to have some selective meaning. The calves of the wildebeests were abandoned, but the desertions were frequently not altogether accidental. These calves were often the slowest to follow their mothers, the last to react to danger, the most likely to wander from the herd, and thus always the most vulnerable. Their deaths seemed part of a grand design by which the quality of life was constantly monitored. When a newborn creature did not possess the necessary temper, the necessary strength, it was destroyed so that it would not become an impediment to those more deserving of survival.

If death seemed totally dispassionate, its effect was leavened somewhat by the ingenuity of those who survived. The young of one female antelope suckled at the teats of another. Young gazelles born one hundred days before, but still in the presence of their fawning mothers, raced forward to jackals attacking the newborn.

All the grazing animals depended for their survival on the capacity of the young to walk almost immediately after birth; the wildebeest calves galloped with their mothers only thirty minutes after they were born. But in those agonizing minutes when strength was pumping into their limbs, getting them, wobbly, to their feet in full view of hyenas and jackals, no lives on the plains were more vulnerable. The zebras, the oryxes, the kongonis, the topis kept moving in search of better grazing while the females dropped their young, forcing these animals to make concessions so that birth and travel together would be possible.

A cow kongoni gave birth to a sickly creature and picked the youngster up in her mouth, trying to drag it with her as she moved, knowing that movement was essential. She lost her calf to hyenas and became a foster mother to two calves who had lost their mothers to lions. Infinite variations on the theme of survival touched every part of the grasslands and woodlands. Topi cows protected four calves while the leader of the herd took the rest of the mothers to the water hole. Young elands gathered around a solitary cow with no other eland visible anywhere.

Because there seemed to be no limit to the variations in behavior now, birth on the plains became a parade of the unexpected and the bizarre, the fascinating and the tragic. The baboon was stopped by a group of oryxes, some of them trailing calves. He did not fear the grazing animals but retreated when one of them advanced toward him. She came within a dozen steps, then lay down in front of him, completely relaxed, while the others continued to graze. She closed her eyes and her flanks moved steadily.

While the baboon remained still and apprehensive, the oryx's birth pangs began. The youngster's head appeared and the oryx stood up so that the youngster could slip out. The umbilical cord was quickly severed, and mother and youngster returned to the herd. In other times, the grazing animals had come close to his troop, bearing their young in full view of the baboons, and he had not caught the symbiotic truth of the act: the birth performed securely amid the finest warning system on the plains or woodlands—the watchful eyes of the sentinel baboons.

A buffalo calf strayed from its grazing mother to drink at a pool and aroused an old cow elephant, who followed it. She seized the calf with her trunk, violently upended it, and smashed it to the ground. Other members of the elephant family appeared and began drinking so that when the alarmed buffalo cow began her charge to scatter the elephants away from her whimpering calf, she was so well outnumbered that her charge was futile. One of the elephants parried her fierce head-on rush with his tusks and sent her sprawling into the mud. As she struggled to her feet, the old elephant cow was upending the calf again and throwing it to the ground. Then, with the buffalo cow's second charge just begun, the cow elephant raised her lethal foot to stamp the calf flat. This time, the buffalo's charge had some effect. She missed being impaled but struck one of the tusks diagonally, pushing the elephant slightly off balance. Puzzled, the elephant cow turned away and was battered in the rear end by a third charge. All the elephants lumbered off together.

Now, the calf was immobilized in the mud, driven into it, practically buried. The cow buffalo lowered her horns, dug them into the mud, and reared, head up, to lift the calf. Eventually, she was successful but the calf could scarcely walk. It bleated pain as it followed its mother obediently, back legs wobbling, and the two animals returned to the place of grazing.

On his third day of travel, the baboon came to the first large group of trees on the plains. They were clustered along a watercourse, running a line of symmetric green across his path. The marabous had gathered there in hundreds to breed, and before he reached the trees, the baboon could hear a familiar sound; it was the cry of young birds calling for food. He was a bird pirate and he hastened forward. Inside the cool green shade of the trees, the marabous snapped their beaks together with loud, clappering noises. Silent during all of their nonbreeding year, they cried out now like stricken lion cubs, grunted and groaned like wildebeests, and the melange of extraordinary sounds made the shadowed acacias a faintly sinister place. As the baboon climbed toward the nearest nest, several marabous took off clumsily from the branches above him, clattering their beaks together rapidly. In the canopy of the trees, he sat easily and ate, surrounded by angry, clumsy, noisy birds.

The young fueled much of the energy of plains life and the baboon would eat enough here to send him a score of miles to the north. The outpouring of small bird eggs alone was influential everywhere. The egg-eating snakes had fared poorly during the nonbreeding season. They had relied on finding the eggs of some of the ground-nesting birds who bred before the rains, or the eggs of those migrants who had come south from the rainy midsection of the continent to breed during the dry season on the plains. But these were difficult nests to find, and a snake might go for a month or more without eating. Now was the time for debauch, for a banquet far beyond the capacity of all the egg eaters to finish.

The snakes worked their ways slowly up into trees, ignoring the small and angry forms around them, so that each tree buzzed and screamed with outraged weavers and finches whenever a snake chose to hunt in daylight. The baboon, stuporous with marabou flesh, sleepily watched one snake moving steadily and deliberately aloft in a nearby tree in which clustered the hanging bags of weavers. The snake was only interested in fresh-laid eggs. Each time he put his blunt snout into the neat hole of a nest, the delicate sensors in the tips of his forked tongue told him whether the eggs had been incubated or not. If the embryos had begun to develop, he withdrew his head and moved on to the next nest. But if the eggs were fresh, he swallowed them slowly, one by one, his almost toothless mouth taking them delicately, never breaking them, slipping them down his long gullet. As they reached the gullet, the snake's muscled body contracted gently, just enough to crack the egg. The vertebrae of the snake's back actually protruded into the gullet to provide the crushing instrument, a necessary tool when the snake was swallowing the larger eggs of ground-nesting birds. As each egg cracked, the snake raised his head; the yolk flowed down past a valve into his stomach and the crumpled shell remained in his gullet. The snake lowered his head and ejected the shell from his mouth. Because this egg diet was such highly concentrated food, the snakes became fat quickly, and this fat would take them through the long lean months when the weavers and finches had gone and the snakes lay half-dormant, awaiting the return of the rains, and with them a new harvest of eggs.

The outpouring eggs brought hungry hunters wherever they were laid. Vultures watched the marabous and sneaked eggs from nests when it was safe. Ostrich eggs bulged white and gleaming in the sun, favorites of both vulture and mongoose, but the eggs remained invulnerable as long as the vulture or the mongoose could not find a stone on the practically stoneless plains. On the shores of the northern rivers, a crocodile buried her eggs in daylight while sharp eyes watched from the air and from the trees, and at dawn of the following morning, the inevitable monitor lizard slipped along the banks of the stream, moving in short, quick dashes as the crocodile eggs waited briefly in the hot sand.

The crocodile never saw the monitor, a reptile himself. He could lie for days on the branches of streamside trees, unmoving and awaiting his chance. The monitor was not fooled by the large area of disturbed sand where the croco-

233

dile had twisted and turned her body. Instead, he darted forward, head up, eyes beaded bright in the new sun just spilling over the bank of the stream, and began digging. He, too, was seen, and his work recognized; no hunter of eggs could be sure of his prize.

A wheeling marabou stork watched the monitor. Ten vultures watched. The monitor only had time to reach down to the top layer of eggs and begin eating the egg mass before he was seized in the pincer grip of the marabou's long beak and flipped aside. The marabou's own sharp eyes examined the surroundings to see whether she had been observed. If she had been seen by the crocodile mother, that creature would come bolting out of the water to protect her eggs.

The marabou waited motionless for an hour, then assuming safety, she reached down and stabbed into the nest. She worked slowly and thoroughly, but the eggs had been deeply buried in layers, and the big stork missed eight of them, all of which would hatch. She finished her feast and flew off, unaware that another feast remained in the sand which other hunters would chance upon later. Other marabous would see the young crocodiles fighting their way into the sun. Their tiny, wriggling, tough-hided lizard bodies would bring the big birds dropping from the sky. As the birds fell, the crocodiles would run with desperation born of their own evolutionary memories of the dangers of land over water.

The first crocodile would reach the water but the second would be seized by the first bird bounding to a landing ahead of it. The massacre then would be complete, the last crocodile scarcely able to break out of his egg and raise his snout from the sand before being seized in a pincer beak.

But there were many nests, and not all would be seen. Many were guarded constantly. Only one survivor was needed, and later he would have more than his fair revenge on unwary storks and lizards.

While the baboon moved north in uncertain fits and starts, the leopard remained trapped on the island. There, he was a silent and somber witness to the animals eddying around the island. He watched the enormous outpouring of calves and the restless movements of the wildebeest young who had been separated from their mothers. The calves ran back and forth searching for the familiar odor, the familiar tongue. But the mothers, with the casualness of their kind, were already miles south of the island, seemingly unaware of their loss.

The leopard watched a gathering of hunters, creatures who worked best by day, creatures he hated with bitter heat. Hyenas slouched toward the milling gatherings of calves, jackals slinking at their heels. Lions thickened everywhere the leopard looked. Only a few hundred feet away, within his grasp, was ample food, but it was inaccessible to him. The calves were ringed by hyenas so bloated they could hardly walk. Some of the young animals were so desperate for food, their senses so misguided by this incomprehensible fiasco, that they tried to follow the hyenas, to accept unfamiliar odors, unfamiliar bodies, and so reach the security of a suck. The leopard dozed while the melancholy cries of the doomed young rang distantly and discomfitingly in his head.

XIII

A lappet-faced vulture waited in steep, rugged rocks so far from the plains he could not see the edge of them. The crags dropped vertically a thousand feet, ending in the tumbled debris of millennia of falls, and the ground swept down through scatterings of shrubs which gradually gave way to thin forest. From this high country, the vulture used his broad wings to capture the updrafts that the endless winds drove against the slope and crag. By launching himself outward, he could be two thousand feet higher than the rocks in seconds, his altitude gain so great that the distant plains came instantly into view, rolling away to a horizon of blue sky and clumps of cloud.

Because of his great size—nearly four feet in height standing erect—the lappet-faced was the dominant vulture, set apart by his overall brown color—the others had white necks and heads—and by folds of bare flesh hanging from his head. No other vulture possessed such a powerful beak; his size and strength meant that he did not always need to feed on the kills of others. He himself could kill, choosing animals already stricken by disease or accident. His prey ranged from small gazelles, occasional bat-eared foxes and small cats, to mole rats and hares unfortunate enough to be caught abroad during daylight hours.

He had no need to concern himself with the black-and-white wedge-tailed vultures, only about half his size, who were usually content to hunt lizards and rats and insects, though they were sometimes present at the larger feasts. The big vulture often saw the wedge-tails hunting birds' eggs and dung beetles. They frequently gathered together in small groups on the open plains near hyena burrows, where they pecked at discarded bones and tore apart the regurgitations of the hyenas. The wedge-tails competed with the hooded vultures, who were not much bigger and were also hunters of insects and small game and scavengers at the fringes of the larger vultures' meals.

These larger vultures were the lappet-faced's most direct rivals, particularly the white-backed vulture which was nearly ten inches shorter than the big brown bird but had a special capacity to reach kills first, even though he was a rarer bird on the plains. Invariably, the lappet-faced reached kills after these white-backed birds, but then, because of his greater power, his kind would usually drive these early arrivals to the edges of the vulture crowd.

The specialist skills of the vultures dictated the order in

237

which they fed and the food they were able to take. The white-backed and the griffon were adept at ripping long lengths of meat directly from the carcass, their heavy skulls and powerful neck muscles giving them this capacity. The lappet-faced and the white-headed were equipped with strong body and leg muscles which they used in conjunction with powerful twisting motions of their heads to tear loose great chunks of flesh, sometimes too large for them to swallow.

When a particularly big carcass was found, the white-backed and the griffon—who both had long necks—were specialists in opening up the carcass and getting inside it, thus preparing the way for other vultures to join them once the dissolution of the body became general. At this point, the lappet-faced's superior strength and aggressiveness were usually enough to drive off the vultures preceding his kind to the kill.

If the larger vultures found meat ahead of the other hunters, the ferocity of their fighting sometimes kept solitary hyenas and lions at bay, while the big birds scrambled over the carcass, squawking, screaming, and lashing at each other with beaks and claws.

As much as any of the plains animals, the vultures were infected by the paroxysm of activity on the grasslands. They came to the plains from far-distant savanna, from valley and mountain, from humid rain forest and rocky slope. The lappet-faced vulture took off from his high position each morning two hours or less after sunrise and ballooned upward in the company of scores of other vultures heading for the same destination. This morning, the skies were empty of clouds and the breeze was steady, so he allowed himself to be taken to the limit of the upward-moving air until he commanded a sovereign's view of his realm from a height of more than twelve thousand feet.

From this height, his sharp eyes saw every twist and turn of life. A flat-topped mountain showed its symmetric shape far to the southeast. The grasslands, toward which he was gliding, expanded now to fill his eyes from horizon to horizon, and he could clearly see the dark lines of moving wildebeests and the open congregations of zebras and gazelles. He drifted so high he was invisible from the ground; yet he could distinguish the tawny blobs of different prides of lions in that ocean of green.

His speed was supercharged by a high-altitude wind moving contrarily west against the eastward-moving ground, an updraft wind. Although he appeared to be gliding effortlessly, he was often nearly out of control because he flew badly. The speeding grip of the wind tended to tilt his broad wings as he hurtled toward his destination, and in the thin, high air, this instability strained pinions and tendons dangerously, and could tumble him. He had seen other vultures caught in such dilemmas; they were transformed into tangles of falling wings, their screams descending to silence. He kept control, though, and his speed was so great that he reached the plains before the sun had had time to warm the air there enough to give the plains vultures updrafts in which to take off.

Nothing flew above him. Even birds more powerful than he usually appeared to him as distant brown wings circling thousands of feet below. The giant bateleur eagle sometimes rose close to his highest altitude, often giving him a display of roll diving which sent the great bird of prey sweeping down to earth at such speed that he diminished to the size of a lark in a few seconds.

239

The high-flying lappet-faced had a seasonal view of life that was unmatched by any other creature. He saw the first rainclouds emptying on the plains and noted the movement of animals toward rain they could not see and grass that had not yet begun to sprout. He witnessed their day-by-day advance across yellow-dry country to a great circular splash of green sponsored by the first rains, and watched their retreat back into the woodlands if the rains did not continue.

When the white storks came under him on their way from the growing winter of the northern hemisphere to their nesting places in the southern part of the continent, they appeared as splashes of white beating steadily across the plains. They flew at one altitude throughout and were not interested in the plains, though they were exceptionally sensitive to birds of prey. The lappet-faced vulture had seen descending eagles create panic among the storks as the great white birds dropped in ragged escape falls, their wings upflung and feet stretched down as though they anticipated collision with the earth. He had often flown above them at his own lofty, thin-aired altitude, and at the sight of him, they invariably collapsed in midflight, plummeting like white stones to lower altitudes.

Pink-backed pelicans flew across his territory with stately grace, flexible necks doubled back into their bodies. They flew high on their way from one lake to another which lay along the floor of a deep valley to the northeast, but the vulture was not a true bird of prey, and he ignored them.

Today, no layers of birds interposed themselves between him and the plains, and so he could see the grazing animals gathered in their manifold nations, mixed yet separate; together, yet apart; collected, yet widely dispersed. Long black lines of wildebeests were trekking to new grazing grounds. Open conglomerations of zebras and bunched groups of gazelles fed placidly. Here and there stood scattered groups of kongonis and topis and thin herds of elands. Small troops of baboons, widely separated and never far from cover, prowled the grasslands as though they were grazing animals.

Under his high scrutiny, four cheetahs who had not fed the previous night were gearing themselves to make a gazelle kill. A pride of lions was hungry and would kill soon. A leopard who had born a cub five nights before was already moving along the track of four baboons who had struck out toward the open grasslands. A solitary baboon was visible far to the north, traveling fast. The vulture marked his position since he represented a likely kill.

Another leopard whom he had often seen now lay motionless in his island tree, an odd place for a leopard since he was so exposed. Near the island, clusters of newborn wildebeests were surrounded by hyenas and jackals, but there were too many competitors for the meat to attract the vulture. At a dozen points, single hyenas and others in groups galloped in search of good killing situations. The cheetahs spread wide, one animal beginning his loping run toward the gazelles, who were standing still, heads erect, as if paralyzed by the approaching danger. They turned away too late, just as the cheetah went into his second stupendous burst of speed. The vulture shifted his gaze to watch a circling eagle some distance to the west, and when he looked down again, the gazelle kill had been accomplished. The vulture turned his wings flat to the face of the wind and began a wide, circling fall, the sound of the wind rushing deafeningly through his feathers.

His falls to earth were always exhilarating, but the speed was so great that he had to watch closely to avoid collisions at lower altitudes. Few birds were able to look up and behind when flying. He had seen one unwary vulture crash into a group of pink-backed pelicans. Two of the pelicans were crippled and fell in untidy masses of feathers. The vulture had been injured and was barely able to hold equilibrium. The lappet-faced vulture had heard the double smacks of crashing pelicans just as the stricken vulture finally lost control and slipped away at increasing speed until he slammed into a shallow pool, throwing up a gout of water.

Less dangerous, but more numerous, were the thousands of smaller birds who flew together: the sand grouse and swifts and flocks of quelea finches. In their overwhelming invasions of the plains, the finches were often so thick that no other bird could fly through their enormous flocks without hitting scores of them. When the vulture descended to a kill, he often heard the finches' massed voices hissing like a high wind through a world of thornbushes.

Now, as he fell, he caught a glimpse of a solitary hyena near the cheetah kill, and he checked himself, considering the other possibilities of the morning hunt. To the northwest, another opportunity was developing. For many days he had been watching an ostrich incubate her eggs on a nest set along the shores of a meandering watercourse. He had a perpetual hunger for the taste of the large eggs, but there was no way he could drive an ostrich from her nest. For that, he needed assistance, and now, it might be available.

The solitary lion whom he had seen the previous day on the southern fringes of the plains had worked his way north during the night and was plodding steadily back toward woodlands country. If he continued his present course, the lion would disturb the ostrich, and if the mood seized him, he might smash the nest. The vulture responded to this possibility by changing his position slightly, allowing himself to plane downwind until he had placed himself midway between gazelle and cheetah, lion and ostrich. There, facing into the gentle wind, he hung in the sky and waited.

The sky above the plains was usually layered with birds, each working from the altitude best designed to guide him quickly to a kill. The vision of each creature was different. Vultures watched for scavenging chances, but some of the smaller ones also looked for migrating insects. A solitary crowned eagle saw a small group of francolins foraging through some thin scrub near a watercourse. He watched a monitor lizard dozing on a branch and some mongooses darting back and forth between a trio of burrows. One of the mongooses dashed off among a scattering of trees pursued by an angry black bird who feared for her nest. The eagle marked this mongoose as the best potential victim. He saw two bat-eared foxes come out of their burrow and look warily around. One remained near the shelter of the burrow while the other trotted off in the direction of a pool of drinking water.

The eagle almost always had a choice of victims, and much of his time was spent evaluating the easiest kill. He knew from experience that the monitor lizard would struggle so furiously that he could not be sure of killing it, much less carrying it. Near the lizard, a tiny warthog followed closely behind a family group and it, too, was a potential victim, perhaps easy to carry off if he could make his stoop without warning the sharp-tusked male leading the group.

The bat-eared fox was by now in an impossibly vulnerable position; there was no hope of his reaching the safety of his burrow before the eagle hit him. But the fox would fight like a mad thing. He would probably make a long, zigzagging run, then throw himself on his back and attempt to fend off the eagle with snapping jaws and flailing feet. This defense would make it difficult for the eagle to get a good, clinching grip, and he might be bitten severely on the legs. He preferred to take a fox from behind when it was in a dead-straight run for the burrow.

This still left the mongoose as the best available victim, since the eagle could see that he was now too far from the burrows to escape. Also, he was transportable. So, with no apparent haste, the great bird turned, downthrusting beak picking up the tones of the wind in a melancholy whistle. He flexed his tail feathers, half closed his wings, and began the long, diagonal fall to earth.

When he was two hundred feet from the ground and a scant second away from the mongoose, the creature saw him. Chattering fear, the little animal turned and raced for the now-distant burrows, but almost immediately realized he could not make it and turned, darting for some nearby undergrowth. The double shift of movement deflected the eagle's attack, and although he struck the mongoose, his terrible talons drawing a scream from his victim's throat, it was not a killing stoop. The eagle paused, covering the mongoose with great, outstretched wings, but he could feel the animal moving in his talons.

The cries of the mongoose brought out his comrades. They chattered and screamed their rage, charging toward the eagle in a tight group. The bird watched them coming, shifting his grip slightly, then took off clumsily, the mongoose's body bumping along the ground while he gained air speed. Behind him raced the screaming mob of mongooses who seemed to be trying to bring the eagle to earth by the intensity of their rage. He ignored them and reached the branch of a nearby tree. He disengaged a talon from the chest of the mongoose to get a hold on the branch and felt the mongoose turn in the grip of the other talon and sink his teeth into the eagle's leg. Wings flapping clumsily, the eagle swiftly transferred his grip to the struggling animal's head, but this still did not kill the mongoose. By an agile contortion of his slender body, he managed to bite the eagle's other leg. In pain now, the eagle let go so that he could refasten the talons, but instead he lost the mongoose, who flipped, turned end over end, and dropped among his comrades.

For a moment, the eagle did not move. The stricken mongoose was bleeding from a dozen holes in his hide. His back legs were partially paralyzed. The concerned family group surrounded him protectively and slowly made its way back toward the burrows. One mongoose remained behind and when the eagle took off to return to his point of vigil in the sky, the mongoose leaped at him. He was a score of feet short of his target but the intensity of his determination was manifest.

Whenever a crowned eagle came to the plains, the lappet-faced vulture kept him in view, because the eagle, like the cheetah, had difficulty in holding his meal once he had made the kill. The crowned eagle was smaller than the martial, and he enjoyed eating the conspicuous colobus monkeys who occupied some of the tree lines along the fringes of the plains. He would spend hours watching them work their way through the trees until, in their characteristic fashion, they took to the treetops to sun themselves. Although several of them might be on guard duty, the crowned eagle had a chance then of putting the sun behind him and coming down at such speed that he remained invisible until it was too late.

After the mongoose fiasco, he circled again at a lower altitude, turning toward monkey country near a trio of islands surrounding a water hole. There, he knew, he would find the white-mantled forms of the motionless monkeys. But he had to be careful because the islands were leopard territory and even his sharp eyes could rarely find a leopard, no matter how long he stared into the cool green tops of the trees. Besides, the monkeys were large victims in relation to his own size, and he had to plan for the aftermath of the kill, for the disposal of the body.

The lappet-faced vulture allowed himself to be transported up, and up again, until he passed out of sight of any creature on land. He remembered the number of times the crowned eagle had lost his kill to jackals, who could then be driven off by vultures of his size. He lost interest in ostrich eggs and the gazelle carcass now being devoured by the solitary hyena.

The eagle was strong enough to kill almost any large monkey, many small antelopes, and newborn gazelles, but his wings were not powerful enough to carry the animals away. He had to dismember them, a clumsy business which meant leaving one part of the carcass behind while he transported the other part to a safer place.

Today, the colobus monkeys were in conspicuous view, stupefied by the warmth of the sun, and the eagle began his stoop immediately. Instantly, the vulture came down from his high place in a spiraling fall that ate up thousands of feet in a minute. By the time the eagle struck among the trees, the vulture was low enough to see that the kill had been successful.

The eagle gorged himself on the intestines, then set to work on the spinal column in an attempt to split the animal in half. He ripped at the vertebrae with his beak, both feet holding the body down while he strained backward and forward. When the break came, he fell backward, then sprawled forward in his haste to get at the severed head and rib-cage section which he could just carry. He transported his booty down the line of trees toward thicker cover.

The weight of the load was too much, however, and long before he reached his destination, he was forced to stop in a low tree. Awkwardly, he draped his kill over a branch close to the trunk and headed back toward the remainder of the carcass. Watchful and tense, he swung into the acacia and crouched over the monkey's torso. Nothing moved around him. The sighing of vulture's wings whispered in the air high above him, but no sound of baboon or vervet or colobus warned him of the presence of a leopard. He gripped the hindquarters firmly and lifted, but realized immediately it was too heavy to fly with. He turned it over and began working away at the belly portion, ripping off strips of skin until he exposed bone. Then he drove his beak strokes deeper until he had a grip on one of the leg bones. Wings erect and flapping, the big bird strained talons and neck until finally his beak found the hip joint, pierced the gristle, and severed it. Heavy-winged, he took off again. He did not see the leopard move in behind him, or know that he had lost the rest of the monkey. When he reached the low tree, his meat was gone.

After the vulture had finished the last of the half monkey,

he resumed his place in the sky. The monkey meal had not been much more than a morsel, and he was still hungry. Fresh meat was a rarity for him. The competition on the plains was so severe that he was usually forced to eat rotting meat, not by preference, but because his appetite gave him no options. Fresh meat disappeared quickly. Hyenas killed and ate mostly at night and never left food into the daylight hours. Lions remained on a big kill for days, and it was almost impossible to steal meat from them. Some lions were so irascible that they made a special point of killing vultures. Wild dogs never left anything except bones, and the hyenas usually claimed those. Only the selective cheetah offered him fresh meat scavenging, and then only when there were no hyenas present to contest the food.

He turned widely across the plains, possessing some sense, absent in other creatures, which could divine distress from signs quite invisible to other eyes. Earlier in the year, he had been present at the birth of an elephant. The vultures had begun to gather in the trees around it almost at the moment it rose on wobbly legs. The mother had trumpeted her rage and dashed at the vultures, shaking the branches of the trees where they waited, but when the young elephant had died two days later, she had slipped away silently into the woods, leaving its body to the birds.

The vulture could identify distress in mature creatures by tiny variations in the normal pattern of their lives. When he spotted a young impala skulking alone in undergrowth, he descended immediately. The impala, its back leg broken during an unsuccessful leopard attack, had hidden as long as possible but now was forced to seek water. He came stumbling out into the harsh sun, the vulture moving abreast of him, and then he fell. He tried to get up and failed. His thick tongue protruded from his open mouth and his breathing was shallow and fast. His appearance in the open brought down other vultures until more than a score of the big birds circled him. At once, the lappet-faced vulture's perception of distress sharpened and he landed close to the injured impala. The helpless animal looked at him with expressionless brown eyes, but he lacked the strength, or even the will, to move. The vulture reached forward and with a powerful movement of his formidable beak, began to rip off strips from the animal's hide. The antelope sank into the shock state of the grazer about to die. He lay with his eyes closed, but still alive, still aware. In seconds, scarcely a part of his body remained uncovered as vultures smothered him. The flesh was succulent, the hide thin, and the press of vultures hid the nature of their work. At the last moment, the head of the dying impala rose above the jostling vultures, eyes staring blank and wide at nothing, then it fell back into the press of feathered bodies.

But the vultures were not to be allowed to enjoy their kill unmolested. The old lion, hungry and irritable, padded toward them, his guttural roars ordering them to leave. In the haste of his departure, the lappet-faced vulture lost the chunk of meat he had just ripped from the fresh-killed carcass. The lion, his authority established, settled down to feed alone.

Still hungry, the vulture turned south, passing across serried black lines of creatures migrating, creatures calving, creatures resting. He ignored the other species of vulture hovering over birthplaces where it was their role, and that of jackals and other smaller hunters, to clean up the placental remains of birth. Before him, the open country gave way to scattered scrub, undulations of the land, and the other vultures disappeared behind him. This was better country for hunting eagles and large owls who needed roosting cover for their work. Ahead, he saw a pair of soaring bateleurs, the largest and most powerful eagles on earth. They were not established on the plains and never would be, but for the moment, they were a majestic and beautiful addition to the aerial spectacle of the vulture's world.

They turned into the sun, the lustrous blacks and reds and chestnuts of their plumage catching the brilliant light. One of the bateleurs abruptly folded his wings half-backward and dropped to the ground in a diagonal flight that was so fast the sound of the wind ripping through his feathers came clearly to the vulture. Looking beyond the eagle, the vulture saw the prospective victim, a puff adder curled on a rock. The snake heard the sinister roar of the eagle's approach, and moments before the strike, it slithered under the rock and the eagle pulled out of his dive.

The vulture sailed above a hunting territory of eagle owls, creatures of both day and night whose booming cries sent shivers through many ground birds. These owls were big enough and powerful enough to influence other birds of prey. They hunted young hawks, still unfledged, and when the vulture was roosting on low ground, he often heard night-blinded hawks trying to defend their nestlings from the owls. He listened to adults screaming as they plunged through the trees and heard young hawks being pursued by the eagle owls who killed them on the ground or while they were hanging upside down from thicket branches.

The eagle owls caught and killed lanner falcons and chanting goshawks and other large birds of prey. Their night forays often intruded into the sleeping and nesting territories of the large eagles, and it was customary for the bateleurs, the martials, and the crowned eagles to retaliate during the day. When they flushed eagle owls in daylight, they chased them with leisurely confidence in their own superior speed and strength, and sometimes killed them far out on the open plains.

The vulture soared on as killer watched killer. A large eagle carrying the remains of an antelope found himself followed by two other large eagles who were determined to pirate what he had killed. In such aerial battles, the ferocity of the conflict frequently became so intense that smaller eagles and hawks and kites became unwitting victors, dropping down to claim the victim when it fell to earth.

In any long flight, the vulture covered thousands of square miles and he always memorized developing food chances. He saw an eland killed by a small pride of lions and knew it was likely to be available to vultures when the lions became satiated. He stored this knowledge. At the far eastern fringe of the plains an elephant had died days before but because of its size, no scavengers had been able to gain access to its body. The carcass had been investigated by lions, by jackals and by hyenas, but the tough flesh had resisted all their efforts to breach it. The vultures roosting in the surrounding trees had pecked out the eyes, but this still did not give them entry into the body.

The lappet-faced vulture turned toward the elephant carcass and when he arrived, he saw many others waiting around the huge corpse. He ignored the inquisitive jackals, a pair of bat-eared foxes from a nearby burrow, and other vultures circling overhead, and landed near the rear of the elephant. The vulture had lived long enough to know something about elephants. He began ripping tiny pieces of flesh from the edges of the anal orifice until he had worked his way beyond the fringe of tough hide to the softer flesh inside. His success brought down two other vultures who attempted to join him but were unsuccessful because the hole he had made was only large enough for the entry of one vulture's head at a time.

By late afternoon, however, fifty of the big birds were working shoulder to shoulder to rip open the entire rear end of the elephant. Some of the birds half disappeared inside the carcass in their efforts to get at the softer and more succulent parts of the elephant's interior. The lappet-faced vulture shared a common frenzy with the other birds; the knowledge that the feast might be interrupted at any time by the arrival of a larger scavenger, perhaps a lion who would drive off the vultures. Their digging had now made the elephant available to practically every meat eater, and the exclusive banquet could not possibly last much longer. Still, they feasted until evening, when the bloated birds staggered away from the body and lurched for the trees. None of them could fly. The lappet-faced vulture thrashed in the undergrowth, clawing his way upward until he was uneasily astride a mass of light branches. He heard other vultures clumsily beating their way into cover until the night fell still around him.

At dawn on the following day, the vultures were back at work. Now, the eating speeded up. More than a dozen of the big birds could get inside the elephant at once, eating their way through the coiled intestines. By noon, thirty of the birds were inside the carcass and the lappet-faced vulture had eaten most of the liver.

When the local pride of lions returned in the mid-afternoon for another look at the elephant carcass, nearly all the vultures were inside and feeding. Only some of the smaller and more specialized birds still waited outside for the time when they would have a chance to snatch scraps of meat. The lions drove off these birds and one of the juveniles poked his head inside the hole opened by the vultures. Instant chaos. Behind him, the lappet-faced vulture heard a concatenation of screeching voices as the vultures there found themselves trapped, and panicked. One of them attempted to make a precipitous escape and hurled himself forward into the face of the lion who crushed him with one bite. But the vulture's hysterical charge forced the lion to retreat a step or two and when a second vulture came hurtling out, he contented himself with a swipe that sent the big bird spinning a score of feet away, bruised and winded, but able to escape.

Inside the elephant, panic gripped all the birds. The lappet-faced vulture was pushed back against the deepest part of the body cavity by the thrust of struggling birds behind him. He could barely move. The flailing of wings and the high-pitched whistling screams of the trapped birds turned the banquet into a nightmare. Meanwhile, the lions sat on their haunches outside and watched as the exit of the vultures began. Like some horrendous cloud of clumsy, angular insects, they came bolting in disorder out of their escape hatch, bird piled on top of bird, feathers stained with the debris of their feast, necks and beaks all green and bloody. Last out, the lappet-faced vulture scuttled among the lions who ignored him as they watched the larger meal ahead of them. He could not fly but he managed to get back into the undergrowth. Then, his desperation fed by the shock of his escape from the elephant, he managed to claw himself up the trunk of a tree from which place he would get a clear view of the end of the elephant.

With the carcass empty of birds, the lions began their own feast. This mountain of meat was too great for them to hold to themselves and so their attitude toward the other scavengers was different from that displayed over kills of zebra and wildebeest. The hyenas came and fed while the lions slept off their first great meal. The jackals worked among the hyenas as the skin and hide weakened under the onset of bacteria. The lappet-faced vulture recovered from his meal and descended during intermissions so that on occasion he was fighting for flesh among hyenas, jackals, young lions, and masses of other vultures.

The elephant was gradually stripped down, laid bare bit by bit under the jaws of millions of creatures—fly larvae, beetles, ants, bacteria, storks, all competing with the larger animals for a share of the meal. The great rib cage was exposed, and the skeleton of the creature appeared. When the lappet-faced vulture finally left, the skull was in view and folds of leathery skin lay coiled into fantastic shapes. The ground around the remains of the elephant was churned and tracked by the movements of the feasting multitude. The big brown bird lofted himself into the warm air, soaring higher and higher until he was gone from sight.

XIV

When the old lion reached the island, the pain in his leg did not trouble him as much as before. He lay down with a groan near a pool of water, licked the throbbing pad of the paw, and looked dully around him before he drank. In these days, everything he saw, everything he experienced, could only conjure memories of other, and better, times. The sounds of lions mating had disturbed him the previous night by their sheer insistence. The insane caterwauling had gone on throughout the hours of darkness as the pair copulated again and again, and this had brought back to him persistent memories of pride life and accented the aloneness of his present situation.

His discontent was increased by the mating he saw all around him. He had witnessed cheetahs racing back and forth in their extraordinary courting dances which ended with one female mating in turn with four or five males. He had sourly watched ostriches perform the long rituals which led to mating, meeting in an absurd sprawl of wings and legs and writhing necks which made it impossible to see what was happening to whom. At his feet, a small tortoise, about one quarter the size of his mate, grunted and groaned as he struggled vainly to mount the back of her smooth carapace, his determination undiminished by the fact that he had spent all the previous day trying, and would work all night, grunting his frustration at failure, before finally achieving his aim.

Mating stirred the lion's juices, but he knew, now, that he would never again possess a pride, that no lioness would pause to hear his roar. He could only slide backward into that time when his power had created a pride and then had expanded it by flinging off progeny to all parts of the plains and woodlands. Of the four thousand lions living there, he had sired several score of the surviving animals, many of them now possessors of their own prides. His power had gone from him, but it still lived on the plains and in the woodlands.

He watched other males fighting bitterly for their places in the hot sun, and sometimes dying in the attempt. Even among the antelopes, the fights could be lethal, an odd twist of fate for creatures who were unable to defend themselves against the meat eaters. The cheetahs never killed in resolving their sexual ascendancy, but the lion knew about the ferocious fights of the leopards, and he had eaten leopards killed in such fights. The hyenas might fight, but deaths were rare in their matriarchal system. The hunting dogs sometimes drove off submissive females in their sexual conflicts, ate the puppies of rival bitches, and generally displayed a viciousness that contrasted sharply with the rest of their cooperative ethic. The lions fought, and how well he remembered that, but if they ever killed each other it was more by accident than from the desire to kill.

His memories were not sequential. They did not measure his history in terms of what had happened, but were expressed rather in the form of pains and pleasures, highlights in long gray periods. He had no memory, for instance, of a two-year period when his great pride had born nineteen cubs of whom seventeen had died. The death of the cubs, one of the main population control factors in lion lives, simply did not register. He had no memory of his own selfishness during the good times when he would commandeer a new kill for himself. He had a blurred memory of bad times when his youngsters had died, nasally meowing at the dried teats of his mates. He had no memory of times when food was short and he had eaten his own cubs to assuage his hunger.

All he really needed was a precise memory of immediate events. He traveled constantly through a series of loosely delineated territories which might, or might not, be defended. To reach the island, he had walked in a curving, southeasterly direction which had taken him out of the woodlands where he had driven the vultures from their impala kill, and over the now-shallow river. His journey had been well watered by scattered showers, and myriad new flowers had blossomed underfoot all the way. He had taken a zebra from an old female hyena after a brief scuffle and had remained on the kill for several days, gorging him-

257

self. When he had been driven from the carcass by thirst —he always drank more deeply when he was eating heavily—he had turned south again and in the early morning of the tenth day, had surprised a leopard on the scattered remains of a gazelle, the leopard unaccountably far from cover, with no tree in sight anywhere. The lion had watched as the bounding cat disappeared.

Always, the old lion suffered from the lack of the hunting female. In the development of his kind, she had been his crutch, his killing agent. Her ruthless determination, her speed and power and her skill in hunting all exceeded his own, and she thus became an extension of his own life. Gloomily, he raised his head, eyes blank, and looked at a pair of antelope bucks fighting each other, their absurd little tails wagging frantically, their butting, short charges not doing any damage. He knew they were fighting for a female. He groaned and closed his eyes. Ten days before, he had encountered a lioness who had aroused a flicker of hope in him that perhaps he might start another pride or at least enjoy cooperative hunting once again. He had seen a herd of wildebeests on the other side of a small rock island, the wildebeests close enough to the island to give the hunter perfect concealment before the final charge. At the same moment, he had noticed the lioness positioning herself near the island. She was crouched in longer grass where she was almost invisible, and while the lion had watched, she had slid forward rapidly, hardly bothering to stalk at all.

Instantly, the lion had moved forward in a flanking movement which would put him on the other side of the island. He had been responding to the ancient lion custom of working both ends against the middle of the hunt, a device that may have been intentional or accidental, but which worked more frequently than any other method. The lioness, hidden from him then, had uncoiled her tense body slowly and charged.

Almost immediately, her attempt had aborted. The power of that first strong kick had been too great for the traction she had in the wet grass, and one hind foot had slipped instead of gripping. Lacking power, she had swerved to one side and galloped away as though the last thought in her head was to kill a wildebeest. The wildebeests had turned immediately in flight, heads down, long hair streaming, and they had thundered straight toward the old lion hiding in his ambush place.

He had been in an easy kill position, but he had been inhibited, as always, by his memories of disaster. He had remembered how frequently young lions were injured in their first naive charges and how some of them were killed by victims who seemed to be aware that the hunter, himself, was fallible. He intuitively understood that his kind were naturally clumsy hunters who frequently bungled their kills, and the image of his old body facing these charging maniacs held him fastened to the grass for several seconds as he strained for the will to move.

When he had acted, it had been as though he were a pride leader once again; a smooth, clean leap took him diagonally onto the shoulder of one of the wildebeests. The force of his charge was not in competition with the impetus of the fleeing animal, as a frontal collision would have been, but complemented its power so that the flying weight of the wildebeest helped to send the creature to the ground. The lion had stayed on top, teeth buried deeply into neck muscles while the animal flailed beneath him. In this position, the lion knew from experience, it was merely a matter of time before he would be able to exchange his grip for the more lethal facial grip that suffocated the victim. Growling his pleasure, he had caught sight of another figure coming from behind him. He had turned his head as much as he could and had seen the lioness. Then, numbed, astounded, outraged, he had felt the crush of her jaws on his shoulder, the thud of her paws striking home to grip his body. In the fight, he had been so severely punished that he had been unable to intervene while the lioness ate his wildebeest.

The drive of other males to assert what the lion had so completely lost went on continuously throughout the woodlands and plains. The bushbucks fought for hours, churning the earth to dust in dry places, until one rammed his horn deep into the chest of the other. The kongonis fought and a loser's leg snapped in one furious affray that left him to die alone. The waterbucks charged each other again and again until both were dizzy with the crashing of horns and skulls. Weaker-willed animals wandered off, the losers. In the woodlands, magnificent kudus used their great horns like spears, but the fighting was more psychological than physical. The statuesque fighting of the giraffes went on in slow motion until one of the great animals went down, perhaps to be kicked to death by the victor.

The fleet impalas resolved their mating conflicts through an intricate fighting preamble which ended with one buck dominating a large harem of up to forty does. Because the bulk of the males were segregated into their own herds while the successful male shepherded his does to graze, these bachelors established their own orders of dominance so that when the opportunity came to challenge a harem owner, there would already be victors of their own sequential fighting, or demonstrating, to take up the challenge. This procedure, ancient as the earth around them, seemed based on mysterious dances, on sudden and often unresolved fighting, with the tension growing steadily among the many thousands of males. Sometimes, the bachelor herds broke into bursts of activity, a dozen bucks chasing each other in circles while three or four others fought. Sometimes, they dashed through the scrub, nose to tail, tails fully erect, and disappeared from sight. Deep emotion gripped all of them during these rituals, the bucks grunting like pigs, their breath hissing sharply through their nostrils.

The resolution of male dominance, of sexual assertion, was never designed in simple forms. In bachelor-herd fighting, two contenders might vie for the chance to challenge the dominant buck with the harem. He, however, might decide to enter the conflict himself and cause such confusion that both the other animals would be put to flight. This denouement broke their fighting urge and sent them chastened back to their bachelor herds. It might be days, perhaps even an entire season, before two more bucks would be ready to resolve who should attack the dominant animal again for control of the does.

The successful bucks, holders of harems, were beset by their own problems, which were well understood by the old lion who often hunted them. He had killed two of them in his brief swing through the woodlands before returning to the plains, both kills the result of the impalas' inability to escape him rather than by any special skill on his part. He knew their exhaustion, their reluctance to run when they sensed his ambush, and he acted on these signs, though he knew nothing of their reasons. The successful bucks had placed themselves in positions of such vulnerability that it seemed a slip in the evolutionary scheme of things. Their does responded only to their constant guardianship and had to be shepherded every hour of the day and night. Each harem had its quiet escort of bachelor males seeking for the quick opportunity that would mate them with one of the protected does. Each harem's buck was forced to patrol his flock almost constantly to foil such attempts.

But the bucks also had to graze and mate with each of the does the moment she came into season. Some bucks had little time for anything other than fighting, chasing, and copulating. With large herds, the buck might have to mate with five or six of his does every day, and fight off a dozen attacks as well. Time and again, when he was mounting a receptive doe, other does were chased away from the herd by opportunistic bachelor bucks and he would be forced to go after them when his mating had been completed. The lion knew only the end result of this frenzy: exhausted impala bucks with threadbare coats and ribs sticking through their hides, so dulled by the frantic pace of their lives that they cared little whether the lion was waiting for them or not. They fell under his paws and lay limp, the fight gone at this final affront to their ambitions.

On the plains, the old lion witnessed other problems of sex and territory. When the main wildebeest herd reached the island once more, the black bull was as exhausted as any impala buck. He had rounded up his family following the hunting-dog attack and the slaughter of the youngsters by hyenas, and as usual, the herd had been almost constantly on the move. This month was his only chance to impregnate his cows, the only period when they came into season. Therefore, his work had a desperate urgency about it, since

262

his success or failure would dictate the quality of the rest of the year for his family.

The lion lay somnolent near a plains water hole, now turning dark with teeming growth and with the disturbances caused by so many feet walking its shallows. Lying there, mouth agape in the heat, the lion dozed, just as he had spent long days sleeping away his injuries after the loss of his pride years before. He remembered those pain-filled days, his injuries suppurating, the river dried to a series of dark pools twisting through tall fig trees. Sandpipers had walked past his nose and flown quickly across the pool, almost like large insects in the intensity of their movements. Above them, small bee eaters had dashed low across the water and disappeared into thickets. He had looked across the still waters and watched baboon and impala come together, united in their common need for water, saw giraffe and warthog walk silently to the edge of the pool and drink cautiously.

He had recovered slowly among the streaming thousands who had come to drink. The water had pulled them out of the dryness of the woodlands in small groups and one by one, but always it was a steady attraction. Dimly, the lion had seen the paw print of a leopard, the drag mark of a crocodile, the pads of hyenas, and he had heard the barks of baboons as he made his long journey through and beyond his private world of pain.

The time of mating across the plains, woodlands, valleys, marshes, and mountains brought all the males to tumultuous climax in their need to find and conquer females. This affirmation of life, this positive declaration of faith in the future, was made in violence, the air resounding to the shrieks and snorts, the grunts and groans, the metallic clickings and clackings, gasps and squeals, roars and rumbles of countless impassioned males attempting to reach the summit of their season.

Because the superior male must triumph over his adversary as part of the process of natural selection, the mating act bespoke not just the life to come, but also the possibility of death. The rhinos' fighting was conducted with grotesque squealings and guttural snortings in front of the desirable cow, turning procreation into comedy until the savage thrusts of their great horns inflicted grievous injuries. Horns were rammed into chests, rib cages, stomachs, hindquarters, the blood flowing so copiously that the two contesting males slipped and slid in it as they gasped with the expenditure of their vital forces.

Eventually, the vanquished male turned and ran, or fell on his side, the last goring just too much for him, to be rolled ignominiously down a bank into a river or pond where he drowned or floundered his way to the opposite shore.

The leopard, tense with his own mating urges, tried to sleep in his island tree but instead became a reluctant witness to the contest between a male and female rhino as the cow, aroused by her season in an odd way, turned on the bull and gored him. So severely was he hurt that he fled, but his place was taken immediately by a new and more passionate bull. The leopard dozed to the squeals of the gored bull and the grunts of the mating couple.

The leopard should have been among the buffaloes near trees and water holes, streams and rivers. There, he would have witnessed their fighting, their savage butting with broad, heavy heads that left many of the great black creatures with broken legs. Such injured animals were often victims of prowling lions and hyenas, or died slowly of starvation. Sometimes, bulls locked horns in their fights and died together, making the nights and days noisy with their rage, then frustration, then anguish at their mutual view of the future.

In his night wanderings, the leopard's acute embarrassment at being among wildebeests and zebras, gazelles and kongonis, was hardly allayed by his occasional contacts with familiar animals like rhinos and elands, topis and wart-hogs. He was more accustomed in this season to the furious fights of the hippos, who made the nights resound with their sexual energy. The great power of their jaws made them formidable fighters, able to kill their sexual opponents, which they did occasionally, or hurt them so badly that the leopard used to hear them staggering through undergrowth to the refuge of their pools, then the sounds of them drowning as they sank into the water.

Mating took so many forms that it constantly presented new faces to old situations. When a cow rhinoceros came out of a thicket with her calf, she faced the ancient problem of many mothers on the plains: how to break the bond to her youngster so that she could resume her sexual life. Her calf had become relatively enormous, but he was still dependent on his mother and beginning to be jealous of her sexuality. The hot sun splashed her dusty hide as she followed her predictable path toward her next grazing ground, her next lavatory. There, at the edge of the plains, she was accosted by two bulls. Immediately, both began demonstrating before her.

The cow was not yet fully in season and she looked at the demonstrations without responding, but the young rhino sensed the significance and he could not contain himself.

265

Enraged, he charged the courting bulls despite the disparity in size. They stood fast and might have killed him if the mother had not charged the bulls only paces behind her youngster. The collision of the four animals was a bizarre accident; none really wanted to fight the others, but all were forced into a comic melee. The young rhino ineffectually butted the bulls with his horn while they tried to parade before his mother, intending no harm either to cow or calf. The ground turned to dust under the pounding feet, and inexplicably, both bulls fled. The slower of them was severely gored in his hindquarters as the enraged female caught him on her horn.

The old lion slept off his wildebeest-calf meal, and now the music of the plains changed again. Instead of the monotonous droning of wildebeests, he heard the gasping hoots of the zebras. In this rutting time, the zebra stallion maintained his rigid authority over his family group, and this mastery meant that he had to fight competing stallions with all his strength and guile. Under his discipline, the members of his group cooperated to help each other more than any of the other grazers, and there was concern for the lost, the stricken, and the sick. The stallion did not physically lead his family, he walked parallel to it or at its rear, since the serious attacks invariably came from behind. His lead mare went ahead of the group, followed in order of dominance by the other mares and then by the youngsters. The family was reinforced by the capacity of each member to recognize the other by scent and voice and at a distance by the patterns of individual stripes, which were distinctive for every zebra on the plains.

In rut, the competition within the zebra herds was not much less chaotic than among the wildebeests. Any stallion who had been successful enough to gather a large family was forced to fight many times harder to keep it than those stallions who had only two or three mares. Unfortunately, at this critical moment, the zebra stallion had nearly all his mares in heat at the same time, and when he looked up from grazing, he saw his group surrounded by enthusiastic stallions, each eager for a chance to mount one of his mares.

He charged forward immediately at such audacity, but in the middle of the fight he saw one of his mares being mounted by a youngster. He rushed to kick him away, only to see two more stallions rising on other mares and a group of them being chased into the distance by several stallions. He ran after them and reclaimed the mares, but his family was in disorder on his return. Mares were being mounted everywhere. He chased away his rivals with bites and kicks.

When the competing stallions were temporarily put to flight, he was left exhausted. In his worst moments, he was incapable of serving mares who importuned him, and when he saw new groups of hopeful young stallions approaching, he felt his age as a weight in his bones and gut. One year, he would be unable to stand this pace and he would be retired from his family. The law of the mating season applied to all those confident males who so thoroughly dominated their females and their families. Their fate was preordained, without hope of change or mercy.

The tide of mating animals moved across the plains, ebbing and flowing around the islands and water holes, passing through floods of rain and sunshine. The old lion killed desultorily in exploitation of moments when the male grazers were so engrossed with sex they forgot that lions remained opportunists. The unmated leopard, edgy with sexual frustration, took his chances too and slid among the sleeping groups of animals to kill a calf here, a juvenile there. The old female hyena, still stiff from the kicks she had received in the last zebra kill, felt no urge to mate. The cubs in the woodlands, now approaching independence, would be her last litter. She ran alone this night across the plains, her side stabbing still with the bruises of the kicking. The baboon used the preoccupation with mating as his shield against the dangers of the open plains. By now, he was more than three quarters of his way toward his destination.

XV

Rain was in the air. Almost every living thing felt its imminence, though no creature had the capacity to measure the quantity about to fall. Confirming the anticipation of the animals, the rain began gently, forming puddles, then slowly grew heavier, turning the puddles into small ponds. It increased in intensity, pounding the earth until it thundered full-throated from one end of the plains to the other.

At first, the baboon had welcomed the rain because he knew it inhibited his enemies. It made the lions grumpy and miserable and disinclined to hunt. Both the dogs and hyenas also preferred not to hunt in wet weather, and even the cheetahs were contained by the rain. But as the baboon loped along toward his destination, he began to feel a dread of the unknown. He had never seen rains like these before. The water fell in almost solid masses around him and blinded him. He dashed the water off his protruding brows, but it made little difference. His eyes filled with water. He put his paws to his head and felt the water teeming down the hair of his limbs. He stumbled along, the water deeper underfoot every second until it was nearly up to his knees.

Because the rains were falling widely across both woodlands and plains, no creature escaped them. Back on his island, the leopard was driven from his tree into an island cave where he forced out a skulking striped hyena. The rains crashed on and did not give the land time enough to absorb the water. Flooding spread. The mole rats, who had been successful breeders in the previous year, were driven out of their burrows, and jackals splashed through partially flooded country, yapping at this unexpected hunting. Tawny eagles arrived and settled in trees near the rats who were swimming or running from the flood. The rain continued, but it did not stop the eagles from planing down gracefully to catch rats everywhere. They waited for jackals to catch the rats, then, with wings outspread, they bluffed some jackals into dropping their catches.

The old female hyena was gripped by the same uncertain feeling of the unknown as the baboon. She splashed through puddles and a flap of skin pounded against her flank. Fresh blood matted her hide there. She ran into the distant curtains of new rain and came to an island where she briefly caught the scent of lion. She recognized the individual's scent; it was the old lion who had nearly killed her at the zebra carcass. She veered away to skirt the island.

Under an overhanging mass of granite, the old lion slept, relieved for a moment from pain throbbing high in his leg.

The pervasive rains went on, seeping deeper into the soft earth everywhere. Birds feasted on insects flushed from the ground. The omnipresent termites faced the rains with varying fortunes. Those who had established their cities well were scarcely troubled, but the rains searched out badly placed colonies, and they were abandoned. For the innumerable new colonies attempting to get a foothold in the grasslands, the rains became a test of mortality when queens drowned or died of exposure. Most young workers, dwarfs in these early generations, could not make the needed adjustment to survive, but adjustment was available to them, and some managed it. With their queens dead, they fed a few of the nymphs in their compact nests with a special food that almost immediately began a transmutation in the nymphs' bodies. The growth of the nymphs quickened, sex organs matured, and although these secondary queens could never be as large as the dead primary queens, they would give the colony an alternative way to survive.

In some colonies, however, no nymphs were developing when the rains began, and when the queen died the workers were faced with a double dilemma. Most failed to solve it and the colonies perished. But a third system of survival was available: the workers could choose some of their own number to become new kings and queens. The workers were creatures whose sexual organs had not developed, and so it was logical that if they were fed a special food to stimulate the development of sexual organs, they could take the places of the dead queens and kings.

In some instances, this was done and a new and third type of queen and consort emerged. It was a compromise arrangement at best because these new kings and queens would develop into the smallest of the many royalty lines available to the termites, and they would retain many of their plebeian origins. They would be blind like the other workers and permanently wingless. These third queens were able to lay fertile eggs but never near the rate of either the primary or secondary queens. Worse, these eggs would not turn into nymphs that could become winged emigrants. Doomed to be workers and warriors, their fiefdoms would remain small and few would survive many years. Buried in this third effort to survive, however, was the idea that into such tertiary colonies a new queen might come some day to complete the transformation, and so set up the permanent foundations for a city instead of a satellite township.

The female hyena ran on, the baboon breasted the currents of the rains, the old lion slept, and the leopard brooded, spray touching his nose as he looked out into the streaming wetness. It might be days before any of them would eat again. Elsewhere, though, the rain represented opportunity, not defeat. The fungus-growing termites were prepared for rain, perhaps because of the exceptional vulnerability of their large queens, who were twenty to thirty times the size of their workers. Indeed, their obesity was such that no passage anywhere in their cities was big enough to admit their bodies, even if they had been capable of crawling, making it vital to protect them in their chambers. The queens lay supine all their lives and did little else except spew forth eggs. Unlike the harvesting termites who had elected to be as inconspicuous as possible and were now reaping the penalties of that decision, the fungus growers' city complexes were towering edifices which jutted from odd parts at the edge of the plains with all the incongruity of structures planted there from another planet. Their cones and domes were visible everywhere, and these vertical buildings needed better ventilating systems, with more control over humidity and temperature, than were demanded by the horizontal communities.

As the rain poured down, the enormous queen lay comfortably in her royal chamber above the rise of water. It seeped into the rooms and corridors below her and sent armies of her workers scurrying higher into the edifice, but their hurried departure did not affect the queen or her tiny, inch-long consort. Her attendants ignored the hammering of the rain which was audible through the thick earthen walls, and continued to lick her body all over, feeding her the special food that made and kept her different from the lowly born members of the city. Other attendants removed the eggs as they oozed slowly from her body.

Despite the rain, workers continued to dig through sodden soil for food; pieces of old grass, antelope excrement, twigs, and leaves. The ground was too wet for them to reach the trees where their main source of food lay, but most of their underground tunnels were aimed at the trees where nearly all the dead wood had been eaten from the inside by their voracious hunting. Whether they worked at the surface of the soil or in a fallen branch, they maintained their invisibility, laboring under the cover of an exterior sheath of wood or under a shroud of earth particles glued together by an adhesive they manufactured themselves.

The rains fell. The workers processed wood, ground it to paste, swallowed it, and expelled the waste matter in their feces. Through some accident or ancient design, they lacked the tiny organisms in their guts that could break down the cellulose in the wood they had eaten, so that the feces contained cellulose, a by-product which was not allowed to go to waste. The feces were collected by other workers and brought together in loose heaps of reddish material which was placed on top of a series of thin clay arches that sheltered the living places of the termites. The material was the genesis of the vital fungus that made the city work. Specks of spherical fungus balls grew there, and these formed the food for the obese queen, for her consort, and for the young creatures being reared inside the city.

In a moment of involuntary mercy, the rains stopped. Sun struck starred reflections of light into new ponds, lakes, and streams. Plains and woodlands steamed in the aftermath of the downpour. The queleas, now in the final stages of their breeding explosion, rebuilt their nests and ignored those that had fallen in the rain. Young crocodiles hugged the bottoms of dashing streams. The carcasses of animals caught by the unexpected speed of the rising waters, including spring hares, the kittens of small cats, young birds driven prematurely from their nests, fawns of gazelles and other antelopes, moved away toward the great lake. The sun rose into landscapes so transformed by the rain that the female hyena stopped on the shores of a plunging river, her crossing denied by the racing waters. The baboon huddled in a lone tree surrounded by water, and the old lion half walked, half swam through a lake that had formed almost overnight near his island.

In the earth itself, the work went on wherever the flood had not penetrated. The warmth brought worker termites to the surface. They hauled material from subterranean fungus beds and spread it out in thin blankets. During the rain's pause, delicate white mushrooms grew from this material, quickly made spores, and died. The fungus grown by the termites was not of a normal kind, having been doctored by some process known only to themselves. The first fungus crops were joined by other fungi, these new types apparently uninvited guests. The termites were not fooled by them and nipped them from the fungus-growing gardens, destroying them.

At midnight on the fifth day of dry weather, the rains began again. The lion took refuge in an island cave and groaned his boredom. The leopard, who was halfway along the familiar route to his hunting grounds, paused while he tried to calculate how heavy the rains would be and whether they would stop his hunting entirely. The female hyena was running upriver to a place shallow enough to ford.

Now, the cumulative despair affected the plains. The lion roared his disapproval, and the fastidious leopard growled and hissed at the empty air. During the first night of the new rains, the buck gazelle felt his limbs gripped by inertia, as though all his senses had become dulled by this stronger force which was beyond his experience or comprehension. The sounds of danger were meaningless to him. He stood unmoving although the whoops of hyenas came through the rain-streaming darkness. He ignored the guttural belch of a nearby lion, and the distinctive whistle of a reedbuck passed into his brain without effect. Around the buck gazelle, other bucks and does and juveniles and fawns were standing with their heads down and ears drooped. Their hides streamed with the endless fall of water. Then further insult was added to the harassment of the rain. Wind rose and slashed the rain among the gazelles. Soon after midnight, the gazelles began to surrender to the storm in universal recognition of the disappearance of their collective will to survive.

As well as killing many of the creatures, the rains disoriented almost all lives. The surviving elephants, only six of them now remaining of the great herd that had come down, anthrax-struck, from the north during the previous year's drought, were as much victims as the smallest birds clinging uneasily to the lashing trees. The old lion, finally free of the tumult of the plains, had reached one of his known refuge places, a shelter among rocks by the river, well covered with a large fig tree. He looked out into the teeming darkness and the pain of his injured leg joined the ache of his hunger. Despite the rain, he scented elephant and understood that they might just as easily scent him. In this shallow refuge, he might be trapped and killed by creatures who hated him so thoroughly.

The rumble of thunder and a sudden, vivid explosion of lightning made the wet air crackle above him, and the lion saw the misted outlines of elephants passing before him. A second bolt of lightning hit the largest elephant, a bull with long, curving tusks, and he went down heavily, his trunk writhing backward in a last contortion, smoke pouring from the back of his head and from his mouth.

The surviving elephants milled uncertainly in the flickering glow of the almost continuous lightning, divining disaster but unable to measure it accurately. They jostled each other in their efforts to lift their stricken comrade. Tusks bit into mud on either side of the dead animal, curved under his body, and with concerted heaves, the elephants attempted to get him to his feet. Others twined their trunks in his and tried to drag him.

The lion could not move. Any attempt to leave his refuge would have put him within sight of the elephants, who were trumpeting and screaming, waving their trunks and stamping their feet in the intensity of their emotions. Young bulls prowled wide of the death scene, trying to find the cause of the disaster. Their screaming roars rose above the sounds of the storm and pierced the lion's ears. He shivered, recognizing uncontrolled passions. A crack, like a tree snapping, came out of the rain, and in a flash of lightning, the lion saw that one of the largest elephants had broken off a tusk in an effort to lift the stricken one. The chunk of ivory spun away in the rain.

For more than an hour, the elephants continued their ritual dance of the dead over the smouldering body. They left occasionally to rampage among the surrounding vegetation, circling the lion's cave and ripping up shrubs and small trees. In periodic flashes of light, the lion saw elephants place leaves over the back of the dead bull until the body was completely obscured. Then, the elephants stood silent around the body. There was a lull in the lightning, and when it flashed again, the lion saw that the elephants had disappeared.

The elephants had gone, but they left behind them such a pervasive stench that the lion felt impelled to move. The smell of the smouldering corpse permeated the air. Ignoring the pain in his foot, he hobbled out of the cave and headed into the full force of the rain. He moved now with uncertain intent through a world only dimly perceived by his strained senses. A rock indicated a place of ambush, a flushing watercourse recalled the edge of the woodlands. He was swinging involuntarily to the east now, moving with a steady, slouching gait, despite his limp, that carried him along steadily. Neither woodlands nor plains now held comfort for him, and he moved in intuitive response to the need to travel. Eventually, in a long and aimless movement, he reached the country of the stricken gazelles. Their over-powering scent came to him in the grip of the wind and rain, and told him they were packed closely together. He took a score of cautious steps, the scent growing stronger, and when he came among them, all their horned heads were turned to him while their rumps faced into the wind. None moved. He paused uncertainly, his hunger sharpening, and then made one clumsy pounce. The gazelle did not struggle and died almost immediately. None of the other gazelles moved; they did not even turn their heads toward the kill. The lion leaped again, and killed again, and again. Some gazelles were so close to his kills that he almost touched them as he sprang. They started briefly to their feet, snorted, moved off a few paces, but then lay down again.

The lion rushed forward, inflamed paw forgotten, and killed and crippled indiscriminately. Gazelles all around him rose and fled together, but even in flight, they still lacked the will to survive and they blundered into each other, even charging into the lion's red jaws as he, panting now with the effort, continued killing.

The rain stupefied. Dazed birds, no longer able to fly, were buffeted in their tree refuges by the chill winds and fell toward the hissing earth to become victims of surface hunters who could work in the rain. The lappet-faced vulture was so well soaked in his treetop refuge that it might be two or three days after the rains stopped before he could fly again. Of all the plains creatures, the hyenas were best able to move freely through the rain, exploiting the misery it was causing.

The old hyena had returned to her den and fed her cubs, then had crossed the river again to hunt the plains. She joined others and they moved together, splashing through new ponds and streams until they came upon the numbed gazelles. They struck at them along the northern edge of the flock, while the old lion was massacring it in the south.

The old female remembered other times when her tribe plunged into such bloodbaths. It was the ancient tradition of practically all meat eaters to kill beyond their needs when easy opportunities came. The killing urge could not be made selective. The kill was the aim, and no power arbitrated the numbers to be killed. The old female's powerful jaws gripped and wrenched, and gripped again, and the whistles and screams of stricken gazelles sounded from the corners of her mouth and from either side of her. The hyenas killed and killed because they were unable to stop as long as this fresh meat continued to lie before them.

The killing was systematic, spread out across a front as broad as the hyenas could make it, each animal working at the flanks of the next, and methodically killing every supine gazelle. None paused to eat. They lunged and jerked and ripped and broke rib cages and crippled and killed, and the rain kept pouring down into a night world made violent and senseless by the scope of the killing.

The buck gazelle was one of the few survivors of the carnage. The determination which had served him so well in other seasons surfaced now under the pressure of the hyena attack, and strength returned to his limbs. Snorting his terror, he zigzagged like a hare chased by a cat and fled from the bloody massacre. Presently, other tribes of hyenas gathered to bloat themselves with food.

When the rains stopped, a long silence stole over the plains. The dead lay where they had fallen. The lappet-faced vulture took to the air, his wings unexpectedly dried in the intense heat that had followed the rains. He came on

279

the gazelle massacre where the old lion lay asleep in the middle of the torn and ripped bodies. Jackals scurried around him. Word of the kill was spreading rapidly. The fall of each vulture was marked, and even as he fell, the lappet-faced vulture saw lines of marching lion prides moving toward the arena where the gazelles had died. Food would unite many of the hunters for the next few days while the gazelles were disposed of. The long moment of pause continued through hot days, with the skies clear and hard and brilliant. The plains were acquiring that waiting atmosphere they had possessed earlier in the year. The rains had so thoroughly altered the face of the landscape that new opportunities, new changes in life seemed imminent.

By the time the water had subsided around the baboon, he was weak with hunger and exhausted by the nervous tension of sitting for so many days in his exposed position. But gradually, the water receded, and gradually the rhythms of life reestablished themselves in patterns that he could recognize, enabling him to move once more. He did not know how far he had traveled, and therefore could not understand that the border country where the woodlands began was a scant two days' journey. But in the territory he must cross lay some of the barest of all plains land, unmarked for miles by tree or shrub, island or hillock or ridge. It was a favorite hunting territory of the cheetahs and of the dogs, a fine place for fast hunters to run down their victims. In his weakened condition, the baboon must cross this dangerous territory by daylight.

Meanwhile, far to the south, another group of travelers had begun its move for the same destination as the baboon sought. Long before the rains had begun on the plains in the early part of the year, heavy rains had fallen to the south. They had aroused a population of locusts which had been relatively dormant for several years. Its range was a narrow belt of land along a southern river, but this had been a year of unprecedented rains, and the locusts had responded accordingly. With the soil soft and moist everywhere, the females had easily dug millions of eggs into the ground. Everywhere, lush vegetation had sprouted to feed the locusts, and the food was so abundant that the millions of freshly hatched locusts had had plenty to eat. Generation had followed generation. With each new generation, slight changes in physical form had occurred in recognition of the fact that a specialist, migrant-type locust was needed for the big flight that now lay just ahead.

The stimulus of the rains had created a growing momentum of movement among the locusts and had driven them slightly insane. They had taken short, exploratory flights, allowing themselves to feel the wind in their wings briefly before falling back to earth. The frenzy, once begun, could not be stopped. Early one morning, several hundred million of them had taken to the air and headed north. This was no purposeful movement in search of one destination. Rather, it was a blind explosion directed outward and controlled by the wind. If the wind had been blowing south, the locusts would have gone with it. And an eastern wind would have taken them into the sea as it had done many times before in previous centuries. But this flight was wind-directed north. Each locust traveled with his rasping jaws working from side to side whenever he landed to gobble down more than his own weight in food every day. The mass of creatures moved into the vision of the baboon at midday of his first day of dangerous travel through the flat, empty land toward the woodlands. They appeared as a dark cloud which spread over the plains. He stood erect, shading his eyes from the sun with one paw, and watched the cloud intently.

The blind movement of the locusts contained its own timing mechanism which was calculated to give them the best chance to cover as much territory as possible. To reach the plains, they had crossed the low, scrub-covered mountains which separated the plains from the southern lands. They had slowed their body processes so they could make this journey to their next source of food. They arrived on the plains while the land was still steaming from the effects of the rains, still moist and springing new green growth everywhere. The baboon stood transfixed as the locusts came down to eat and to deposit their eggs.

The locusts had made no independent decision to land on the plains, though the new growth there would have drawn them down. But an end to the wind had compacted them into a dense mass, and with the wind gone, they had been forced to land. The baboon could see them streaming down to the plains like black rain a few miles away where a large flock of wildebeests was trudging south in search of less waterlogged grazing grounds.

The baboon saw the wildebeest heads come up, one by one, the big animals ignorant of the existence of these creatures who came to the plains only once or twice every hundred years. The noise of the locusts was a soft susurration in the ears of the baboon, but it was a sibilant rushing of heavy winds to the wildebeests, with a metallic hammering in the background. The black-maned antelopes turned to flee, but it was like trying to escape a rainstorm. Whichever way they looked, locusts were falling in thickening streams, the sound of their clattering wings rising to an unbelievable uproar.

The baboon saw the wildebeests scatter, and then they disappeared into the black deluge. Some of them escaped, only to be engulfed by the spreading stain of falling insects. Beyond his vision, the wildebeests bolted in every direction. As quickly as one group got flight momentum going, it crashed into another group racing away in the opposite direction. The gasps and squeals and honks and roars of the animals rose above the steady rumble of descending locusts. Confusion increased as the ground became slippery with trampled locusts pounded into a glutinous mixture of insect and earth.

While the baboon watched, still transfixed, the noise of the approaching locusts increased. The sky darkened. Insects hit him, bounced, then took to the air again. The sound of the locusts striking the ground deepened into a soft thunder of falling bodies. Locusts clung to his hair, crawled into his ears and mouth, smothered his eyes. The baboon panicked. He ran, and fell, and was immediately covered with the crawling insects while others rained down around him. The ground under his feet moved with insects and he had the panic feeling they might suffocate him. He pushed ahead blindly.

Meanwhile, the wildebeests, who were never calm in crisis, had turned into maniacs. They trampled calves to death and trod each other into cripples. They stomped in the sticky muck that was the debris of the locust landing. Fallen wildebeests suffocated as locusts were sucked into nostrils and lungs. Locusts crawled over bellies and backs, locusts were in ears and eyes.

Eventually, the panic was ordered into one direction of

flight. The groaning mass of surviving wildebeests turned south. The flight took force, thousands of the trampled animals rising to join the charge. Not all the wildebeests were able to leave though; remaining behind were hundreds of bawling calves, their bodies crawling with busy insects.

For the locusts, none of this was important. Even as the baboon ran, they were spreading out. The deaths of a few million of their number under the hooves of the wildebeests did not noticeably diminish their ranks. They landed and began to eat. The baboon's panic diminished as he saw them busy. Before the day was done, he had become oblivious to almost everything except to gorge himself on locusts.

In this way, he dallied among the locusts, finding night refuge in the trees of a nearby island, while the insects grazed the ground to bare earth and deposited their eggs. He stayed to eat as the male locusts moved on in waves and the eggs hatched and the young hoppers emerged. They awaited a favorable wind and took off and the baboon was left alone in the middle of devastation. No blade of grass showed anywhere. The trees and shrubs of his refuge island were reduced to skeletons. The ground creatures—the mice and hares and ants and termites—prowled earth empty of food. The baboon headed north again, and kept moving until the breath roared in his lungs and his limbs became heavy.

None of the locusts had ever traveled this journey before, but they were bound to continue until the energy of their movement subsided and they found themselves in a country where they could settle. In fortunate years, this might be on continents thousands of miles beyond the one in which they now flew.

Ahead of them lay a broad belt of mountain country where food was scarce and nights were cold. The stony ground would not permit them to start new generations of migrants. Unless the wind strengthened, or there were unexpected rains in this hostile region, the host of locusts would perish. The last of them disappeared, and the sound of their wings hung in the air above the bones and skins of dead wildebeests, the footprints of investigating hyenas, the faintest flushing of new grass struggling upward to restore the plains behind them.

XVI

Long before dawn, the lion felt a new flush of pain in his leg. It was like a hot thorn, driven up into his shoulder and heading for his heart. The pain was so intense that he tried to get up and so remove his body as a target of the agony. As soon as his paw touched the earth, his whole leg turned into a haze of unbearable pain and he responded with a roar which died quickly to a strangled gurgle. He had no power to think. But he did understand most clearly that with a paw useless, he would not be able to kill. This was no flash of insight; it was a slowly growing comprehension that perhaps this hunger that had plagued him now for days would eventually make him weaker, and weaker.

The brain of the lion went no further. Groaning, he lay down again and rested the blazing paw, trying to sleep.

The rains had flooded and killed and caused many involuntary migrations, but in a landscape so eager for moisture, so well adjusted to exploiting every drop of water, the aftermath of the rains was a prosperity far greater than the temporary devastation. Within days, more grass had sprouted on the plains than all the animals combined could ever hope to graze, and the grasses made their rush for posterity by thrusting up countless flowers and seeds that would insure their survival into the following year. The calves and fawns of the wandering antelopes prospered in the midst of such grazing wealth; the success of locusts was imitated by almost every other insect on the plains and these teeming new creatures became the genesis for the success of larger animals in the chain of life. The hunters easily found victims from among those made fat and heedless by the abundance of food.

The lion had slept for two days after his gazelle meal, and then had killed again—a zebra this time—before returning among the sprouting grasses at the shelter of the island. The inflamed paw, which he had then been able to subordinate to the pleasure of gorging himself stuporous, had remained a dull, heavy throb that went clear up his left leg.

He had suffered infections before, but somehow this inflammation had weakened the whole leg so that in the critical moment of the kill when his back legs were broadly splayed to get maximum grip and his claws dug into the earth, his left front paw had difficulty in striking that crippling blow that would bring down the animal. It was weak. And because he was left-pawed, it was difficult for him to force his right paw to do what his left paw had always done.

Despite the rain, he remained totally a creature of the habitual movements and moods that his age imposed upon him. It was too late now for him to make an emergency adaptation to the crisis of the crippled foot. Had he been younger, he might have sought a place of refuge in the mud swamps along the shores of the great lake and earned a slender existence hunting frogs and lizards and mice. But age imposed its own imperatives. Instead, he lay for several days breathing slowly.

The lion's dilemma passed into an infinity of enigmatic survivals and deaths, commonplace on the plains. The seasons moved, insensitive to the needs of the animals. The rains fell and the droughts rolled on arbitrarily. The plains did not change essentially. There, chance was seized and turned into immortality amid such sharp changes of mood that chaos and cosmos, abundance and poverty, drought and wet, all lay inseparably close together.

The hairline division between opposites made the miracle of the plains and its ever-changing panoply of the unexpected. Because the torrential rains had fallen at an unaccustomed time, right at the end of the season of rains, they had distorted the fabric of life everywhere. Birds were breeding late, or early, and the grazing animals had fattened when they should have been turning toward drier pastures. It was, however, all ephemera. After the torrent, the glass-blue sky turned clear and remained so, and there was no hint of rain in the crystalline firmament.

The vulture saw the first signs of the change. With the earth now well dried, and the grass growth slowed or halted, the herds of grazing animals were forced to move more quickly from place to place to ensure their exploitation of the last of the sweet grasses. He flew over the network of tracks which marked the plains like the veins in a leaf. The veins formed a series of interweaving patterns that went to every horizon. Along these trudged the wildebeests, now coalescing into larger and larger herds, some of them stretched for miles under the vulture's wings. They moved in long, strung-out multitudes, or spread half a mile wide in scores of thousands.

He flew over zebras, more widely spaced and collected into clots of animals, but using the same trails in their trudging migration to the next pasture that might lie over the horizon. He flew over the recumbent lion, asleep in the open country near the island. He flew over scatterings of the smaller gazelles who did not respond as precisely as the wildebeests and zebras to this changing of the season because, unlike these other grazers, they could go for long periods without water. The larger, stripeless gazelles, who seemed to need no water, would remain on the plains and subsist on the last scraps of food, dry or green, left by the others.

This was the beginning of the drought. The sky had emptied itself of rain, and distant sea winds which had brought the moisture changed their directions; not a single cloud reached beyond the mountains to fly over the plains. With the juvenile drought came the first whisperings of dry winds, quite different from the humid winds of the rainy season.

The first of these winds came in the early evening, the quiet, hot day dying in a flush of blood in the western sky, the sun passing down through multihued layers of color. It was so sharply different from any other breeze, so intransigently dry and hot, that it instantly awakened the old lion and set off new premonitions. He got to his feet and limped back toward the island. The pain of the paw had now immobilized the limb, telling him to rest, to sleep, to give his body a chance to fight the poisons spreading inside him. But the wind told him he must move. He knew all about the difficulties of the sharp transformation from lush growth to drought. He did not welcome such rapid change, nor did any other animal. Tradition, as well as age, made him conservative, and he needed time to move, time to consider, time to fit into the new mode of life. He did not want an overnight disappearance of most of the grazing animals on the plains, as sometimes happened when drought descended suddenly, driving the animals away in helter-skelter retreat. As a nomad, he would have to hasten after them, which was anathema to his lazy cat nature, since he then had to work as hard as the simple animals he killed. Worse, when his victims moved north, they dispersed and so became increasingly more difficult to find.

Indecisive, he limped painfully under the lea of the great pile of rock. The sight of the upthrusting body of the island, forbidding in the purple light of the late sun, offered him no food, and with a groan, he lay down again. Almost at once a liquid growl sounded from the middle of the darkening air above him.

The leopard, for so long the silent and patient watcher of the lion's distress, the observer of his restless pacings, his sleepless agonies, his lurching visits to the water hole, had by now built up such a pressure of frustration that he was desperate for resolution. His hatred of lions coalesced into a single shaft of venom aimed at the solitary and vulnerable lion. Now, in a final and insulting act, the lion lay between him and his route to the north. Like the lion, the leopard suffered the effects of the slow disappearance of food animals. He was just as responsive to the message the hot, dry wind was bringing him. His hunger, on top of this, made him all the more savage.

The lion lay wakeful and still as the light faded into night. As a pride animal, he had regarded the leopard with tolerant contempt, a skulking cat who could sometimes be forced to give up his prey, and incautious enough to store his kills in trees where the lion sometimes reached them. But as a nomad, the leopard had become a deadly competitor for the same food, an irritating and intolerable presence in the lion's own country. He heard the leopard make his decision, listened to the scuff of claw on bark, the chatter of foliage shaking, and the cushioned thump of twin paws touching the earth.

The lion turned his flank to the island, expecting the leopard to come down the steep rock face directly at him, but when he did not immediately appear from among the trees, the lion turned his head toward the island. In that moment, he saw the leopard streak silently down the rocks to the floor of the plains and head straight for him. The lion came half-erect, back paws splayed to get maximum grip in his characteristic lion-fighting stance. The leopard, however, rarely fought so frontally, and he slipped under the broad paws, slamming into the lion's vulnerable chest and belly region. He bit and ripped the inside of the lion's thigh, and spun clear before the bigger animal came down on top of him.

Now, all pain forgotten, all frustrations overwhelmed by this decisive call to action, the lion was rejuvenated. He charged, half caught the leopard slipping to the left, his blunt claws digging for a moment into the leopard's hide, but just as quickly the leopard's teeth bit his paw to the

291

bone. The lion barely felt the wound, so hot was his temper, and he was unaware of the explosion of pus and blood from the injury.

Only later, long after the leopard had gone and the lion was cleaning himself, did he realize that his paw had changed. Much of the pain had gone. He licked it, tasting its corruption. His tongue reached deeply into a fissure in the flesh from which pus still oozed. For the first time in many days, he let out a rumbling growl of satisfaction. With his foot on the grass, now cleansed and pain-free, he sniffed the night air with a hope as fresh as when he had been a juvenile.

The beginning of the drought, touched now with briefly fierce winds, sucked dry thousands of shallow ponds, dropped the levels of water holes, and dictated radical changes in a multitude of different life forms. The kingfishers, harvesting their fish crops along the banks of diminishing rivers, would soon dart away, pausing at water holes before moving across the dry scrublands of the north to isolated lakes, to northern rivers, and so into the center of the continent, where new rains would give them sustained hunting.

The nightjars were preparing to make the same journey but at a different tempo. Their travel would be made at night across purple-black skies with the plains and woodlands ghostly gray under them and dotted with the black marks of millions of slow-moving animals. The storks would fly high and heavy as they headed for their northern marshes, which were already filling with the overflow of a long river that fed the central-west section of the continent.

If the lion had any real premonition of the meaning of this season, in relationship to his condition, it existed only in the deepest recesses of his brain. He did what he must to survive one day. The flights of birds to wetter worlds were meaningless marks in the sky, signifying nothing to him. The arrival of other birds who liked the drought and who would feast on the countless seeds now scattered throughout woodland and plain had no impact on anything he did. The diminishing howls of jackals at night, rearranging their territories as their prey moved from the plains, told him nothing. He cared nothing for the disappearance of vultures from the skies, all of them now gone with the grazing animals to the north.

He took one trudging step after another, himself diminished in an empty landscape of yellowed green. Unlike the leopard, so bound by his territory, he was free to move where he wished. He was led on his limping way by the image of a refuge about two days' journey to the north, where a line of trees marched across the plains, and a creek bed retained water deep into the drought in a series of pools. There, sandpipers ran along muddy shores and swallows hawked across the greenish water. There, he would be among barbets dropping down to drying grasses while rollers plundered teeming termite cities near him. Geese would burst from out of shaded pools, while weavers stood in nearby trees and reedbuck bolted from dense growths of bamboo.

But the memories of the lion were distorted by his age, and he fashioned them to conform to what he wanted. He imagined catfish making sluggish roils in the muddy shallows and an incautious baboon watching for a catfish meal rather than a hungry lion. The shrunken river had been a place where white-headed eagles flew and where the queleas came in their millions as they, too, retreated from the plains, appearing first as a thin stream of dustlike material in the clear air before thickening into a single, dense mass of life, an uproar of tiny finches whose collective weight bent the branches of the acacias. Once watered, they took off with a throbbing roar.

The lion endured lingering pain to reach this specific place where emotion, not correct memory, told him he might survive. He remembered that the streambed meandered with glints of dark green water showing among the lighter green leaves. This was truly a place of pause, where the trees buzzed and piped with the sounds of birds. Each branch and leaf, each pool of water, each speck of earth was patrolled by myriad creatures treading in each other's tracks. In one day, a patch of dusty ground hosted chats and starlings, cordon bleus and whydahs, lizards and bustards, quails and termites, hunting spiders and tiny antelopes, barbets and hyenas. The earth spilled out a black eruption of termites in response to a careless antelope's foot. Larks and storks, curlews and mice, hares and grasshoppers were followed or flanked by a host of bugs, ants, and beetles. Layers of vultures laced the sky to top an interlocking system of life that diminished the lion to a speck of insignificant matter. Toward this dream, this notion, this flawed memory, he trudged on, and the flies followed him closely.

The leopard, in his flight from the island and the lion, widely skirted the headquarters of the dogs, who had settled in the rolling country west of the island. But on that night, the warm wind had brought him a much diminished scent of the dogs, and he divined that they had left their camping ground. It was normal for them to move their headquarters when the grazing animals migrated north.

But the leopard was not to know that his persecutors had moved away in outright flight rather than in orderly migration.

The dogs migrated in much the same manner as the grazing animals, although the line they took was more direct and less susceptible to chance divergences from the main route. They remained victims of a tradition so ancient as to be beyond recall; the most feared of all the plains hunters, among the most prolific, with litters of thirteen and fourteen pups, the dogs still moved to the accompaniment of disaster.

In their complex hierarchies and subhierarchies of male and female, pups were frequently slaughtered to satisfy the desires of the dominant over the submissive, and there was mortality enough to exert some control over their numbers. With their almost flawless hunting techniques and their fecundity, they should have been the dominant killers everywhere, but instead were rarities on the plains. As they moved now, they were caught in the glow of the bursting new sun, a straggle of dogs strung out beyond the waterhole complex. Some turned back repeatedly to sniff at following dogs, and two litters of pups—eighteen youngsters in all—were scarcely old enough to keep up with the adults. Already they were whimpering their distress at the speed of the march. The dogs moved with sickness, and this was the unalterable fact which governed their numbers, preventing them from ever becoming the dominant animals of the plains. The sickness came to them almost automatically every time they experienced prosperity and their numbers soared.

Dull-eyed, tongues lolling, the stricken dogs tramped behind their healthy comrades. The disease, distemper, caused massive gastric hemorrhages and rapidly depleted the dogs, particularly the very young. They became thin and their hindquarters weakened. Some whimpered as they trotted along, and these were dogs which would probably collapse suddenly and die within minutes. The sick exuded gray mucus from their eyes and mouths, and the weakness of their back legs was punctuated by occasional and uncontrollable strong twitches of the muscles which knocked some dogs down.

Almost every relationship within the pack was distorted by the sickness of the dominant dogs, by the attempts of submissive dogs to assert themselves, by the concern of the bitches for their puppies, and by the memory of other death marches, all ending uncertainly in the hot haze of the beckoning horizon.

Toward the end of the second day, two of the puppies had fallen far behind the pack, and their efforts to keep up, signified by whimpering yelps of concern, sent them stumbling over each other as they saw their mothers and comrades growing smaller ahead of them, the scent of the pack diminishing in their nostrils. As the pack disappeared, a jackal came into view. He had been following the dogs in the hope of capitalizing on any kill they might make.

At almost the same moment, a female in the pack went down, muscles twitching, eyes half-glazed, flanks heaving. The dogs paused uncertainly. The dominant dogs were sure of their roles as long as the pack was a unit and healthy. But in such travail, decisions were difficult. During this time, the two puppies caught up with the pack, gamboling among their brothers and sisters with joyful squeaks. The jackal wheeled away and was lost against the sinking western sun. The dominant male lay down, then the dominant female, thus resolving the dilemma by establishing a bivouac for the night.

The dogs exemplified the fragmentation of the life of the plains. Its multitude of disparate parts now fled in slow motion to every horizon, a mélange of wriggling, speeding, soaring, trudging bodies. The baboon, that miraculous survivor, had delayed his own migration to eat among the locusts, and now he faced the last leg of his journey across open grassland that was more thoroughly patrolled by enemies than any other part of the plains. There, in a strip of territory not more than a dozen miles wide, were gathered all the opportunists and prime hunters: the hyena individuals and tribes, the nomadic lions and resident prides, the jackals and vultures following the migrating wildebeests, and finally, a family of four cheetahs who found this ideal country for their lightning-bolt charges because of its lack of cover.

The baboon, of course, understood nothing of this, sensing only that he was near the end of the plains. The locust banquet had revived his determination to move, and to keep moving, until he rejoined his own kind. And so, intent on traveling, he did not see the cheetahs slumped down on a large termite mound, their limp bodies denying the power concealed in them. They looked more like empty pelts flung to earth than lethal cats. One of them saw the baboon, and above, the lappet-faced vulture watched the proximity of baboon and cheetah. He turned into the wind to wait.

When the baboon at last sighted the cheetahs, one of the cats had risen from the mound in a long lean stretch of his body. The uncoiling of body thrust up the incongruously small head so that the animal appeared to be standing almost erect on his back legs, but this was an illusion. His front paws still touched the termite mound. Now, he began his intent, silent appraisal of the baboon, as though he were watching grazing antelopes. The eyes and the head remained absolutely still as he measured all the information he needed to make an attack. His scrutiny brought one of the other cheetahs up with him, and both cats watched the motionless baboon.

Caught naked in such a wide expanse of nothing, the baboon knew enough of cheetahs to understand his dilemma. Before his panic flight from the troop, he had frequently watched the cheetahs hunt from the comfortable and safe vantage point of a tree. He had noted the unvarying ritual of the attack and kill, knew how easily the big cats could be driven from their kills by a determined group of baboons, knew too how keen the competition among the opportunists for the cheetahs' kills.

Above everything else, he knew from his skirmishes with leopards how any running baboon, caught in open ground, was doomed. Slowly, he felt the panic gripping him as the need to run, to flee madly for trees which did not exist, fought with his more sensible impulse to stay still, to outwait the cheetahs, outstare them, and trust that they would tire of the deadly game first. An hour passed. The cheetahs rose, one by one, to make intent appraisals, then sank back to the termite mound again and dozed. This patience of the cheetahs was honed to such a point of perfection that the actual chase and kill became the less significant parts of the hunt by comparison.

The baboon was not habituated to such extended tension; he had no stomach for long waits, and at the end of the hour, his time had come. He turned and began running. It was a blind rush, made sensible only by the fact that he continued to run north. Not even the danger of the cheetahs could head him from his original course. He loped ahead, front legs helping to boost his body forward as they drove into the dry earth under him. He ran without emotion, his panic now restrained by action, by the solace of movement, and for long moments he did not look back to see whether the cheetahs were following him.

Two of them were, in fact, already in motion, running with that high-rocking movement of cheetahs not intent on making an immediate kill but rather of placing themselves in a more strategic position for the final blazing run. The two remaining cheetahs were off the mound, but had not moved far from it. The baboon risked a quick look backward, and the view of the cheetahs revived the panic. The baboon began a frenzied run, his breath booming hollowly in his lungs. Behind his panic, rage lurked, a strong force that would keep him dangerous to the end. It was fueled by the memory of his aching, countless days on the island and by solitude now so prolonged that he could no longer measure it. This rage, which lay deep within every baboon, was the spirit that could kill leopards or mangle hyenas when it got a chance to vent itself. But now, the only release of the pressures inside him was in running, and he stumbled along as fast as he could, half falling ignominiously into abandoned burrows, tripping on low termite mounds and once crashing to the ground when his foot struck a dried chunk of wildebeest dung.

The cheetahs, understanding that their victim was helpless, accelerated to the next stage of their attack, the half-speed gallop that cut down the distance between them and their victim. One cheetah moved slightly ahead of the other; it had apparently been decided that he would make the kill. Even as the separation occurred, the kill cheetah accelerated to full speed, his great bounding leaps eating up the remaining six hundred feet that lay between him and his quarry at a rate of nearly one hundred feet a second. At that point, the cheetah could no more control the outcome of his collision with the baboon than he could make an instant stop. He was programmed by millennia of tradition to reach that supremely coordinated moment when one paw would deftly trip the frantic, running victim. Then, as the cheetah put his four feet down and braked in a long smoking skid, he would seize the tumbling victim and ultimately kill him.

Panic speeded the baboon but he did not lose his wits entirely as would a gazelle or other grazing animal. His brain kept control over a small part of himself and advised him that he should not be a helpless victim. He must finally stake his life on one last bold attempt to save himself. The baboon's long incisor teeth were not meant entirely for tearing up meat; they were also for defense, and as he ran, he turned his head as much as he dared for a glimpse of the cheetah. The big cat was still three bounds away when the baboon pivoted at full speed, so that for one long second he was hurtling backward, his body bowed, front paws flung up and teeth bared fully, facing the cheetah.

The cat, unused to such exceptional behavior, hesitated at the precise moment when his right paw was reaching forward to whip the baboon's legs from under him. The cheetah's attack depended on the victim running from him, not turning to the attack. He braked, body thrown back, but he could not avoid slamming into the baboon's teeth. The baboon tumbled, stunned for a second by the force of impact. But it was the cat's fall that was the more spectacular. His throat ripped by teeth, his long body catapulted over the rolling baboon. His momentum was so great that the baboon was on his feet before the cheetah even hit the ground. The cat landed on his rump and was projected into a second catapulting fall. Even though he had been slowed by his first collision with the earth, this second fall was more serious because he landed with his legs beneath him. Both front legs snapped, his muzzle went to earth, and the third catapult began even as his neck broke and his hind legs came over his head. He landed in a long, skidding fall, dust obscuring the contortions of his body.

In retreat, tragedy was common enough through the plains and woodlands. The steadily moving wildebeests and zebras were themselves shedding their sick, their unfit, their injured, and their old. Now, because the growing drought had turned the migration into something approaching a rout in places, the trail of victims lay strewn behind the moving animals. During the big rains, many wildebeests had been injured in the typical panics of their kind, and these animals, with malformed or badly set broken limbs, hobbled desperately to keep up with their colleagues. Some gathered in the shade of trees and lugubriously watched the passage of the others until they were alone. The retreat seemed to be a time to reclassify the roles of the animals, weeding out those who had proved themselves unfit for the long drought that faced them. Zebras were culled from family groups with an abrupt savagery that left them standing in puzzlement while the group moved on and disappeared. A host of new liaisons formed among the rejected; old zebra stallions and wildebeest bulls consorted in small groups which might cohere throughout the dry season until age picked them off, one by one.

Above the migrants endlessly wheeled a perpetual umbrella of vultures, and with the migrants traveled the inevitable parasites who had never stopped their work from the time the migration south out of the woodlands had begun. The parasites had killed thousands, but now, with the stress on, the death rate speeded up. Some animals lay exhausted, flanks heaving, while their comrades passed on to the north. The animals moved on in their endless, patient lines, heads down, tails switching, to the growing sounds of flies' wings everywhere. It was time again for the warble flies, and they now attacked the rears of the marching animals, laying their eggs around the flanks and legs, while botflies swarmed in the faces of the animals and clustered on their chests. The anthrax, which had been forced into partial quiescence when the animals dispersed onto the plains at the beginning of the rains, flared into new activity as they bunched around water holes or trod in each other's footsteps and dung, mingling their blood in the attacks of the meat eaters.

The wind whistled lonely melodies in thornbushes where the plains had turned yellow with drought and eddied with dust left by the departed. The stench of death lingered in the wake of the migrants. Humped bodies of abandoned victims lay where they had fallen, too many of them to be eaten all at once. At the water holes, growing urgency touched the animals as new waves of migrants overwhelmed the dwindling resources. In the dry season among the permanent water supplies of the woodlands, each animal observed some kind of priority in the order of drinking, even elephant families waiting until groups ahead of them had finished drinking and wallowing. But now, it was every animal for himself in the rushes for water. And when particularly large waves of animals came in from the plains, coalescing at one place, the pools were tromped to mud and bogged animals mired in the muck. At every water hole, flies swarmed, and the sounds of their wings mingled with the voices of the drinking masses.

With the disappearance of the animals from the plains, the old hyena felt a release of the tensions. She had never felt such a pull toward her cubs, themselves never so well placed in such a safe position, never brought to such an age with no losses. They were thriving, and now, making her last foray to the plains, she was returning to the den. Despite her injuries, she galloped like a zebra, rocking along for hour upon hour, her clumsy udder swinging beneath her like a bag. It did not matter that the plains had dried so quickly or that the grazers were leaving so rapidly. Her cubs would become independent as she began hunting in the woodlands.

Now, at last, the animals would be back where she could reach them easily. The survival of her family was assured. She skirted a milling water hole, roaring to the sounds of voices, and passed on into the distance of thickening trees where vultures hung in the sky and the sun swung down quickly at her left flank. It went into a blazing silhouette through the acacias after she had crossed the river, and when she came up and over the last ridge, the familiar watercourse was on her left, a dark patch of green in the drying landscape. She began her final run down into the den. When she was a score of steps from home, she stopped suspiciously. Alien smells came to her. She called, but the den did not answer. Footprints marked the den entrance, dig marks, and closer, a gaping, empty hole.

For ten days, the leopard had held his position in a new

hunting territory, but one now made eerily empty by the absence of animals. He had reached the islands circling the water hole, and no challenging leopard had met him there. The smell of his enemy remained, as did the scent of a female, but both animals had gone. Now, interposing itself between him and his true territory was that daunting expanse of open country where no tree, no shrub, no rock offered refuge. He had little stomach for such a long journey through alien country, but he could not remain amid this group of islands. The trickle of water feeding the hole was drying and the pool would soon darken and become undrinkable. At midday, he looked into the somber distance of the empty plains, delaying decision again, and again.

When hunger finally forced him to make his choice, his desertion of the trees and the islands left him naked. Alone on the open grasslands, he moved uneasily, like all true arboreal cats, half running, half crouching, as if he expected attack at any second even though the grasslands were now empty of almost all life except odd warthogs, occasional jackals, mongooses, and scattered groups of gazelles. He fled north, the day transformed into night and the moon shining high over his shoulder. Shortly before dawn, he reached familiar territory, a place he knew so well that he did not need to see the trees stretching away in a counterbalancing line from a patch of swamp.

He waited, hesitant for a long moment, and as if answering his uncertainty, the cough of another leopard sounded out of the gloom ahead of him, and an answering cry came from further down the line of trees. His territory was still occupied, as it had been ever since he had been driven from it. But now, at least, the decision had been made. His strength had returned. He could not now turn back, never. He moved forward quickly.

In the trees, shaking with new fear, the baboon waited. The end of his journey had been as unexpected, as unpredictable, as its beginning. When he had fled the cheetahs, leaving them puzzled watchers at the death agonies of their stricken comrade, he had known they would not chase him. He had run due north into an emptiness seared white by the sun. The hoofmarks of the departed millions marked his road, and he galloped along their pathways, stumbling over their dried dung, urging himself onward with a desperation sharpened by the nearness of his last escape from death, until he reached the thickening trees.

His meeting there with a troop of his fellows had been a surprise, defeating the expectations he had so long maintained. He had caught the smell of baboon long before he reached the trees, and had heard the squawling cries of a baboon in distress, the cries so emotion-filled that he had increased his speed. But as he had dashed into the dappled shade of the trees, he had found no refuge. The troop was in the grip of some crisis, its organization broken down. The baboon had run among screaming juveniles, been attacked by a young male and then, near the center of the troop's body, had been assailed from all sides by a dozen adult creatures. He had fled, bloody and hurt, to the fringes of the troop.

Like the hunting dogs and so many of the grazing animals, the baboons had been stricken by one of their oldest enemies, a lung infection that lurked in their systems at all times awaiting the signal to act, and then, springing into action, caused them to cough, to choke, and then die of hemorrhaging lungs. The troop had been stressed, then

305

restressed, by the unaccountable rains, by the unpredictable movements of leopards, by the sudden coalitions of
lions, and by days of shivering wetness when they had not
dared venture from the trees. In this time, all the conditions
were satisfied for the outbreak of their lethal disease.

The baboon moved uneasily with the troop, instructed
by the power of the animals in the center to keep his position on the flank. The movement of the troop had been
purposeful, and the baboon, influenced by its mass determination, had not looked behind, or slowed his pace, as the
stricken and the crippled fell. The troop marched on. Sometimes, the baboon could hear the chop and rip of teeth as
the deserted baboons were cut down by hyenas or jackals or
young lions, but he now shared the adversity of the troop,
and lost all comradely feeling for those who had fallen behind.

Now, he looked down fearfully into the dark. The troop
had taken to the only trees suitable for refuge, none of them
particularly high, none possessing those long, slender, high
branches where a baboon might wedge his rump and sleep,
reasonably certain that no leopard could disturb any other
part of the tree without wakening him. This night, however, the baboon was resting close to the ground because
all the high positions had been taken by the dominant
adults and by mothers with their young. He was, in fact,
little more than a single leopard's leap from the ground, and
the smell of leopard surrounded him, rising pungently from
the soil beneath him to linger in the embrace of the trees.

The scent grew stronger, became overpowering, and was
accompanied by a long hiss which ended in a jagged, spitting warning. The baboon understood nothing of leopard
language, but he knew this was a fighting cry, that the

306

stronger scent meant there were two leopards beneath him. He could not know that the two cats faced each other in a corridor running through thick undergrowth, both inflexible in their determination to hold sway in this place.

The rest of the baboon troop was now thoroughly awake. Youngsters clung to their mothers while the elders moved uneasily along thin branches as they tried to estimate whether a leopard attack was imminent.

The baboon was close enough to see the spotted coat of one of the leopards moving forward with infinitesimal caution, the spots on its hide passing a hole in the thicket leaves. Then the thicket exploded.

The screams of the fighting leopards, so close to the troop, sent shock waves running among the baboons. A juvenile who had seen a leopard kill a friend thirty days before made a blind leap into the darkness. Other baboons felt his horror and soon many of them were jumping. The bodies fell past the low-lying baboon, some of them dropping into the thrashing undergrowth, and the panic hit him too. He leaped just as wildly, just as blindly.

When the final crunch of bone sounded in the night and the thicket quietened, the victorious leopard was unseen. He dragged his opponent backwards out into the open. The body lay scarred and ripped under the bright moon. Then, master of his own territory once more, he sat and looked at the silent trees around him, at the moving moon. Somewhere in that gloomy place, he knew, there was a female leopard and soon he would find her as part of the full possession of this place. He smelled baboon, but that did not interest him now, and he did not care about their panicked fragmented flight through the darkness. He would spend the rest of this night savoring his victory, his reconquest of

307

his old land, prowling from one end of it to the other, sniffing old and familiar places. He might indeed meet another leopard, but now he was confident that he would fight and win again, because his strength had returned, his determination was like fire inside him, his terrible sojourn on the island the reason now for his irresistible power over the place that he had won and that he would now keep.

The zebras marched into the drying remains of the great growths of grass that had sprung up throughout the woodlands during the wet, grass which would give them fodder for the coming dry season. They met and mingled with other races of zebras, some of whom had spent the rainy season in the north, in hill country, others who had grazed along the shores of the great lake. The black wildebeest led a straggle of creatures from the plains one morning, the last of his kind to leave, carcasses visible in the emptiness behind him. His head-downward, plodding movement seemed defeated, but in fact, he and the rest of his kind remained unchanged by the disastrous retreat. The buck gazelle, miraculous survivor, remained at the edge of the plains country. His group had thinned out when thousands of the gazelles dispersed, some of them penetrating the long-grass country, others spreading back into the plains, small herds heading slowly for the blue hills of the southwest. They were under lower pressure to seek water, and their close-cropping teeth could find shoots of green grass where no other eyes could see them.

Thus the plains were emptied, or so it seemed, the migratory birds gone into another hemisphere across swamps and deserts, seas and oceans, some of them to the fringes of icecaps. Some insects, like the mosquitoes, already were into their dormant forms, waiting for the next rains. The termites had shed their wings long since and were working to expand or create subterranean cities, freed now from their food-gathering efforts during the growing time.

Soon, within ninety or one hundred days, it would be time for fire. Fierce fires fueled by the quantities of dried vegetation which the grazing animals and the termites and the ants and beetles and other grass-eating animals had been unable to consume would, in fact, become obligatory in many parts of the plains and woodlands, if only to ensure virile growth during the next wet season. In the wake of such superabundance, the fires might be awesome, traveling miles without halt and gripped by seasonal winds that could hurl them forward with devastating strength, or they could burn for twenty days in leisurely consumption of everything: the humus of the soil, the roots of still-living trees, the abodes of insect and reptile and warm-blood alike.

These fires might, in the end, catch the anthrax-ridden survivors of the elephant migration from the north. The elephants, reduced now to a group of only five animals, would be moving back toward the north in response to the call from familiar territories. But the fires would certainly spread, and burn particularly fiercely where elephants had traveled in the previous year, and in years before that, where they had destroyed groves of trees and left dead trunks supine in dead grass. In these places, the elephants might be tempted to linger, and there the flames would burn brighter and higher as they consumed both undergrowth and grass, both living shrub and dead tree. It was not uncommon for such fires to encircle animals, and elephants were not infallible. Perhaps fire was allied to the parasites of the elephants, the only force capable of reducing their great bulks and returning them to the soil. In the smoke and crackling stench of the long drought, the screams of distant elephants would be just another measure of long adjustment to another season, another year.

Fires would certainly drive lions into reluctant flight and fill the air with fleeing winged insects, falcons, flycatchers, and other birds darting through the smoke to catch them. Fires would sweep the woodlands and test the mettle of trees and grasses, of shrubs and herbs, all of which had to resist in manifold ways in order to survive. The bark of some trees was corky, receiving the heat, absorbing it and blackening, but not actually burning.

Some grass would burn to the ground, but the roots would live on. Others produced seeds that were fire-resistant enough to survive the scorching. The fires would suggest death and would ravage landscapes and fell trees, but the devastation would presage the distant season of green when trees would show new leaves before the rains began. The rains would come again and the woodlands would be dotted with orchids and the plains smothered with flowers.

Through the smoke of the fires, the baboons would continue to march, still gripped in the prolonged crisis of their sick lungs, coughing and spitting blood as they moved on in growing urgency to find a territorial homeland. Their flight from the country where the disease had struck them would have the same inevitability about it as the movement of the retreating hunting dogs. But unlike the dogs, whose cooperative ethic was strained by the desertion of any colleague, the baboons would march on regardless. During all this time, the refugee baboon would slowly reassert his authority over one baboon after another, as he worked his way toward the center of power.

The cries of the stricken would have no effect on him. The red line of burning grasses would loom up, and pass, and he would shade his eyes with his hand and look forward, always forward, to where both the unknown and the leaders of the troop lay. The smoke masked a hopeful and successful new season coming inevitably to the baboon beyond the lines of flames.

With the leopard emplaced, the baboons gone, and the grazers dispersed, the old lion lay in peaceful solitude. He had killed and scavenged, and killed again. He had contested the body of a topi caught in the mud of a water hole with a hyena and had sent her spinning with a hard blow that had opened up an old wound on her side. There, with the power returned to his left paw, he had listened to the sounds of the plains. Thin voices spoke of the release of pressure, the disappearance of effort, the flight of discontent. The voices appeared pessimistic, but the lion felt no discouragement. The cycle of the plains had been performed according to the ancient rules, and he had survived.

He had no eye for the old female hyena, disemboweled by his blow, her ugly snout and blind eye raised imploring while distant jackals barked inquiries at the dying moon. Hungry lions ate their cubs and a cheetah starved to death with a small thorn poisoning one paw. The living plains showed that the gentle and the cowardly often survived and the strong died hard deaths; that aggression was rewarded, but only within fixed limits. Everywhere, a fine and incomprehensible balance was at work.

The old lion stood up. A strange heat in his shoulder sharpened into strong pain, and then disappeared. The moon spun and turned red. The abscess in his body which had been the reservoir for some of the poison of his injured paw had burst and its matter was now on its way through his body to his brain. He began trotting, feeling strong as a young lion now, trotting directly south toward the distant island where, faulted memory told him, his best times had been spent. He traveled all night and reached the island at dawn.

His strength grew, encouraged by the reddening morning horizon and he roared at the hidden sun, roared with such power that the phantasmagoric antelopes and horses around him veered away in their thousands. The island was cloaked in green and gray and red, its trees filled with chattering birds, and there, sprawled on the rocks, slumped down in heaps around the shores of the island, lay his pride. He roared his triumph at this moment, roared at the red eye of the sun as it rose and lit him crimson, roared and roared and roared.

Despite the damage caused by the long stay of the baboon, despite the leopard's sojourn there, despite the flood and drought, the hyraxes had recovered, repopulating the island. They foraged on the gray and yellow plains around it. They stood upright, rodentlike, on large rocks, in trees, and looked into the eye of the sun. An old lion had collapsed near the island. He was trying to get to his feet. His mouth kept opening soundlessly. He was obviously dying and helpless. The hyraxes did not see any danger in him.

Instead, they faced into the sun which hissed into the sky and became blazing white. The brilliant, smoking light struck into the crevices of the rounded granite bulk of the island, now seemingly lost again in the middle of the near-lunar landscape around it. Its mass of energy remained unmoving, the impossibility of its escape from its union with earth perpetuated by its interior life: the provision of shelter, the catching and holding of rain, the resistance to the sea of dust around it.

Hyraxes draped themselves over rocks in comradely heaps and let the blessed, life-giving heat sink into them. Dry air lapped at the shores of the island.